Philipp Dietrich

Expansion of the Severe Accident Code MELCOR by Coupling External Models

Karlsruhe Institute of Technology

KIT SCIENTIFIC REPORTS 7726

Expansion of the Severe Accident Code MELCOR by Coupling External Models

by
Philipp Dietrich

KIT Scientific Publishing

Report-Nr. KIT-SR 7726

Dissertation, Karlsruher Institut für Technologie (KIT)
Fakultät für Maschinenbau

Tag der mündlichen Prüfung: 27. Juni 2016
Referent: Prof. Dr. habil. Andreas G. Class
Korreferent: Prof. Dr.-Ing. Xu Cheng

Impressum

KIT Scientific Publishing

Karlsruher Institut für Technologie (KIT)
KIT Scientific Publishing
Straße am Forum 2
D-76131 Karlsruhe

www.ksp.kit.edu

Print on Demand 2017 – Gedruckt auf FSC-zertifiziertem Papier

ISSN 1869-9669
ISBN 978-3-7315-0603-4
DOI: 10.5445/KSP/1000061998

Expansion of the Severe Accident Code MELCOR by Coupling External Models

Zur Erlangung des akademischen Grades
Doktor der Ingenieurwissenschaften
(Dr.-Ing.)

der Fakultät für Maschinenbau
des Karlsruher Institut für Technologie (KIT)

genehmigte
Dissertation
von

Dipl.-Ing. Philipp Dietrich

Tag der mündlichen Prüfung: 27. Juni 2016
Referent: Prof. Dr. habil. Andreas G. Class
Korreferent: Prof. Dr.-Ing. Xu Cheng

“Wer von Anfang an schon sicher weiß wohin sein Weg führen wird,
wird es nicht sehr weit bringen.“

Napoleon Bonaparte

Credits

Die vorliegende Arbeit entstand im Rahmen des vom Bundesministerium für Bildung und Forschung unter dem Förderkennzeichen 02NUK028A-G geförderten WASA-BOSS (Weiterentwicklung und Anwendung von Severe Accident Codes - Bewertung und Optimierung von Störfallmaßnahmen) Projektes am Institut für Kern- und Energietechnik (IKET) des Karlsruher Instituts für Technologie (KIT). Bei der Ausfertigung dieser Arbeit wurde ich von vielen Personen unterstützt, denen ich an dieser Stelle danken möchte.

Herrn Prof. Dr. habil. Andreas Class, Leiter der AREVA Nuclear Professional School (ANPS) am KIT, möchte ich für die Übernahme des Hauptreferates sowie die wissenschaftliche Unterstützung bei der Anfertigung dieser Arbeit besonders danken. Des Weiteren bedanke ich mich bei Herrn Prof. Dr.-Ing. Xu Cheng, Leiter des Instituts für Fusionstechnologie und Reaktortechnik (IFRT) am KIT, für die Übernahme des Koreferats.

Für die Möglichkeit diese Arbeit am IKET in der Gruppe für Unfallanalysen (UNA) anzufertigen, möchte ich mich beim Herrn Prof. Dr.-Ing. Thomas Schulenberg, Institutsleiter des IKET, bedanken.

Herrn Dr.-Ing. Alexei Miassoedov, Leiter der UNA-Gruppe, möchte ich für die organisatorische sowie wissenschaftliche Unterstützung bei der Anfertigung dieser Arbeit bedanken. Ein besonderer Dank geht an Herrn Dipl-Phys Frank Kretzschmar für sein sehr hohes Engagement bei der fachlichen Begleitung dieser Arbeit sowie die vielen hilfreichen Diskussionen.

Meinen Kollegen am IKET möchte ich für die vielen fachlichen und nicht-fachlichen Diskussionen und das sehr angenehme Arbeitsklima danken.

Bei meiner Familie besonders bei meinen Eltern möchte ich mich für die Unterstützung in den vergangenen Jahren herzlich bedanken.

Obersöllbach, 20. August 2016 *Philipp Dietrich*

Abstract (english)

Understanding the behaviour of molten core material in the lower plenum of the *RPV* is essential to correctly predict severe accidents in *NPP*s. Especially for the *IVR* strategy, which aims to stop the progression of a hypothetical sever accident in the *RPV*, a detailed description of the phenomena in the lower plenum is necessary. However, current integral codes often use coarse nodalisations and simplified models, which cannot provide the necessary details to evaluate the applicability of the *IVR*.

There is a variety of new special models, which allow detailed and precise descriptions of specific phenomena in a severe accident with the same computational effort as an integral code due to the improved capabilities of current workstations. The Phase-Change Effective Convectivity Model (*PECM*), as one of these models, was developed based on *CFD*-investigations of molten core material in the lower plenum at the *KTH*. This model uses empirical correlations to calculate the convective heat transfer to solve the energy equation. In comparison to the integral approach in the system code *MELCOR* this model provides a more detailed simulation of the molten material in the same computational time.

However, the *PECM* still cannot capture all heat transfer phenomena in the molten pool. Therefore, an extension of this model was developed to enable the resolution of the stratification in the molten material. Compared to the original *PECM* approach the *S-PECM* shows a significant better agreement with the experimental findings in the *LIVE*-L7v test.

To use its capabilities in severe accident simulations, the *PECM* was coupled to *MELCOR*. Therefore, the *PECM*-methodology was implemented into the *CFD*-software OpenFOAM and then coupled to *MELCOR* using the developed tool *DINAMO*. The validation basis for the *PECM* and the coupled system was expanded by simulating different experiments in the *LIVE*-facility.

The application of the coupled system to the simulation of a severe accident in a generic *NPP* showed the capabilities of the coupling. The new coupling methodology and the extended *PECM* allow for more detailed and realistic simulations of molten core material in the lower head of the *RPV*.

Abstract (german)

Das Verständnis des Verhaltens von geschmolzenem Kernmaterial im unteren Plenum des *RPV* ist erforderlich um einen schweren Störfall korrekt vorherzusagen. Besonders für die *IVR*-Strategie, deren Ziel es ist einen schweren Störfall im *RPV* zu stoppen, ist eine detaillierte Beschreibung der Phänomene im unteren Plenum notwendig. In Integralcodes werden jedoch grobe Nodalisierungen und vereinfachte Modelle verwendet, welche nicht die nötigen Details zur Bewertung der Anwendbarkeit von *IVR* liefern können.

Es gibt eine Vielzahl an Spezialcodes, die eine detailiertere Beschreibung einzelner Phänomene in schweren Störfällen ermöglichen. Das Phase-Change Effective Convectivity Model (*PECM*) wurde auf Basis von *CFD*-Untersuchungen geschmolzenen Kernmaterials im unteren Plenum am *KTH* entwickelt. Dieses Model verwendet empirische Korrelationen um den konvektiven Wärmetransport in der Energieerhaltungsgleichung abzuschätzen. Verglichen mit dem Integralcode *MELCOR* ermöglicht dieses Modell durch die gesteigerte Rechenleistung aktueller Computer eine detaillierte Simulation des geschmolzenen Materials in der selben Rechenzeit.

Mit dem *PECM* ist es jedoch nicht möglich alle Wärmetransport Phänomene im Schmelzepool aufzulösen. Um die Stratifizierung des geschmolzenen Materials darzustellen wurde das Modell erweitert. Im Vergleich zum ursprünglichen *PECM* zeigt das entwickelte *S-PECM* eine deutlich bessere Übereinstimmung mit dem Versuch *LIVE*-L7v.

Um das *PECM* in Störfallsimulationen zu verwenden, wurde eine Programm-Kopplung entwickelt. Dazu wurde das *PECM* in die *CFD*-Software OpenFOAM integriert und über das selbst entwickelte Programm *DINAMO* mit *MELCOR* gekoppelt. Die Validierungsbasis des *PECM* sowie des gekoppelten Systems wurde durch Simulationen verschiedener Experimente in der *LIVE*-Anlage erweitert.

Die Anwendung des gekoppelten Systems auf die Simulation eines Störfalls in einem generischen Kernkraftwerk hat das Potential der Kopplung aufgezeigt. Mit der neuen Kopplungsmethodik sowie dem erweiterten *PECM* sind detailliertere und präzisere Simulationen geschmolzenen Kernmaterials im unteren Plenum möglich.

Publications

This doctoral thesis is based on the following publications:

1. P. Dietrich; “Expansion of the model-basis in MELCOR”; 6th Meeting of the European MELCOR User Group (EMUG), April 2014, Bratislava, Slovakia
2. P. Dietrich, S. Gabriel, F. Kretzschmar, A. Class and A. Miassoedov; “Expansion of the model-basis in MELCOR”; Proceedings of the 45th Annual Meeting on Nuclear Technology, May 2014, Berlin, Germany
3. P. Dietrich; “Expansion of the model-basis in MELCOR”; 7th Meeting of the European MELCOR User Group (EMUG), March 2015, Brussels, Belgium
4. P. Dietrich, F. Kretzschmar, A. Miassoedov and A. Class; “WASA-BOSS: Expansion of the model-basis in MELCOR”; Proceedings of the 46th Annual Meeting on Nuclear Technology, May 2015, Frankfurt, Germany
5. P. Dietrich, F. Kretzschmar, A. Miassoedov, A. Class, W. Villanueva and S. Bechta; “Coupling of MELCOR with the PECM for improved Modelling of a Core Melt in the Lower Plenum”; Proceedings of the 23rd International Conference on Nuclear Engineering, May 2015, Chiba, Japan
6. P. Dietrich, F. Kretzschmar, A. Miassoedov, A. Class, W. Villanueva and S. Bechta; “Extension of the MELCOR Code for Analysis of Late In-Vessel Phase of a Severe Accident”; Proceedings of the 5th International Youth Conference on Energy, May 2015, Pisa Italy
7. P. Dietrich, X. Gaus-Liu, A. Miassoedov, F. Kretzschmar and A. Class; “LIVE Experiments on the In-Vessel Melt Pool Behaviour with external cooling Conditions and the Simulation Results with a coupled MELCOR-PECM Approach”; Proceedings of the 16th International Topical Meeting on Nuclear Reactor Thermalhydraulics, September 2015, Chicago, USA

Acronyms

The following acronyms are used in this thesis:

General acronyms

CFD	Computational Fluid Dynamics
CHF	Critical Heat Flux
HEPA	High Efficiency Particulate Air
IVR	In-Vessel Melt Retention
LIVE	Large Scale Experiments on In-Vessel Melt Relocation and Retention
LOCA	Loss of Coolant Accident
MCCI	Molten Core - Concrete Interaction
MPI	Message-Passing Interface
NPP	Nuclear Power Plant
PSA-2	Probabilistic Safety Analyses of Stage 2
PVM	Parallel Virtual Machine
PWR	Pressurized Water Reactor
RPV	Reactor Pressure Vessel
SAM	Severe Accident Management
SCRAM	Emergency Shutdown of the Nuclear Reactor

Institutions

GRS	Gesellschaft für Anlagen- und Reaktorsicherheit
HZDR	Helmholtz Zentrum Dresden-Rossendorf
IKET	Institute of Nuclear and Energy Technologies
IRSN	Institut de Radioprotecion et de Sûreté Nucléaire

KIT	Karlsruhe Institute of Technology
KTH	Royal Institute of Technology
MIT	Massachusetts Institute of Technology
SNL	Sandia National Laboratories
U.S.NRC	United States Nuclear Regulatory Commission

Programs and Models

DINAMO	Direct Interface for Adding Models
ECCM	Effective Convectivity Conductivity Model
MELCOR	Methods for Estimation of Leakages and Consequences of Releases
PECM	Phase-Change Effective Convectivity Model
S-PECM	Stratified Approach for the Phase-Change Effective Convectivity Model

Program specific functions

CF	Control Function
CV	Control Volume
FL	Flow Path
UDF	User Defined Function

Contents

1 Introduction

Heat transfer from internally heated materials is of interest in numerous fields of current research. In astronomy a lot of examples of heat transfer mechanisms from heated materials to cooled boundaries can be found [1]. The sun in the solar system is one of these examples. The fusion of hydrogen to form helium occuring in the core of the sun creates a tremendous amount of heat leading to temperatures of about 15 million °C in this region. This energy is transferred through the different layers of the sun and is released at the surface as radiation. At the sun's surface the temperature is decreased to approximately 5 500 °C [2]. The outer layers of the sun are heated from the inside by the radiation of the core and are cooled from the outside by the surrounding space. Due to the lower density of the hot material compared to the colder regions at larger radi an unstable state results. The heat transfer by radiation and thermal diffusion is not able to transport enough heat to the outer layers. This leads to the formation of convection zones and the occurrence of the Rayleigh-Bérnad-Convention. The lighter, hotter material is transferred to the surface of the sun, where it is cooled, and sinks back towards the core where the material is heated again. This is a continuous process and the effect of this convection can be seen on the sun's surface where a characteristic pattern of hotter and colder regions can be observed [3], [4].

A similar phenomenon occurs in the earth's mantle. Nuclear reactions within the core heat the earth's layers, which are cooled from the outside at the earth's crust. The cooling from the outside while the material is constantly heated from the inside leads to convection in the liquid magma [5]. Manifestations of these heat transfer phenomena can be found at different locations on the earth. For example, in the middle of the Atlantic Ocean two tectonic plates drift apart as a result of the rise of hot material in this region [6].

The heat transfer from internally heated materials which are cooled from the outside is also of major interest in nuclear safety research [7]. Due to a hypothetical severe accident the radioactive core material can melt and finally relocate to the lower plenum. Examples of such behaviour can be found in the severe accidents in TMI-2 [8], Chernobyl [9] or Fukushima [10]. The severe accident in block 2 of the Three Mile Island Nuclear Power Plant (*NPP*) on March 28th in 1979 showed the risks of nuclear technology. As a result of this severe accident the investigations in severe accident mitigation measures were intensified. The TMI-2 accident is a typical example of a severe accident in a western type Pressurized Water Reactor (*PWR*). A brief overview of this accident follows.

On 28th March 1979 at about 4 pm in the block 2 of the Three Mile Island *NPP* in Harrisburg, Pennsylvania the condensate cleaning pumps failed. The reason for this failure remains unknown. However, the valves to bypass this pump did not open which interrupted the feed water for the steam generator. The interruption of the coolant supply led to an automatic shutdown of the turbine and the electric generator. Furthermore, the valves to feed water into the steam generator by the emergency cooling system could not be opened because they had been mistakenly closed in a previous maintenance.

The missing secondary side cooling led to a rise of the water temperature and pressure in the reactor. Due to the high temperature and pressure the Emergency Shutdown of the Nuclear Reactor (*SCRAM*) was initiated. A pressurizer safety valve opened and released steam from the primary circuit into a tank, which led to a normalization of the temperature and pressure. However, due to a malfunction this safety valve did not close and additional steam was removed from the primary circuit, which led to a further pressure decrease. The operators did not detect the malfunction of the pressurizer safety valve and started the high pressure water injection system to mitigate the loss of pressure.

Due to the lower pressure the water in the reactor began to boil and a bubble of steam formed in the upper head. The increased volume of steam forced the water out of the primary circuit through the open pressurizer safety valve. According to their guidelines the operators shut down the high pressure injection system. The now interrupted coolant supply in the primary circuit permitted the water to continue boiling.

About 1 hour after the failure of the condensate cleaning pumps the main coolant pumps started cavitating due to the increased amount of steam in the primary circuit. At this point the operators shut them down to prevent damage to the pumps. Up to that point the progression of the severe accident could have been stopped. Due to the interrupted coolant support the water left in the core evaporated. The further heating of the core structures damaged the fuel rods, leading to the release of radiation into the containment. Due to these higher radiation levels the operators realized the open pressurizer safety valve and closed it.

174 minutes after the initiating event the operators restarted the main coolant pumps. The water entering the core, which at that point was at a temperature of over 2400 K, instantly evaporated leading to further destruction of the damaged core structures. The destroyed core material melted and formed a pool in the core region. The restarted high pressure injection system was

able to provide sufficient water to re-flood the core region. This formed a crust around the molten material which prevented its relocation to the lower plenum.

The decay heat in the molten material, however, led to a break up of the crust and the relocation of 20 tons of molten core material to the lower plenum. Fortunately, the Reactor Pressure Vessel (*RPV*) did not fail and the molten material was cooled by the water in the primary circuit. The information on the accident progression is taken from [9], [11] and [12] and can be found in more detail there. The TMI-2 accident showed that molten core material can be retained inside the lower head of the *RPV*. However, investigations on the damaged reactor showed that the lower head almost failed in the TMI-2 case. Therefore, a strategy was developed that aims on externally cooling the molten material in the lower plenum through the *RPV*-wall by flooding the reactor cavity with water. The intetion is to to preserve the integrity of the *RPV* and to stop the further progression of the severe accident. This Severe Accident Management (*SAM*)-strategy is called In-Vessel Melt Retention (*IVR*).

After the shut down of the reactor heat is still generated in the core by decay processes of the radioactive material. This decay heat internally heats the degraded materials. Similar to the heat transfer process in the mantles of the sun or the earth, internally heated liquid material is cooled from the outside. If the material is able to relocate to the bottom of the lower plenum, the *IVR*-strategy can be applied. In order to keep the integrity of the *RPV* and prevent the radioactive material from entering the containment building, the cavity surrounding the lower head of the *RPV* is flooded with water. In addition to the cooling of the degraded material through the *RPV*-wall, water can be injected directly into the *RPV* to cool the material from the top. A schematic sketch of this cooling condition and the resulting heat transfer processes in the molten material is shown in figure 1.1.

The lower part the molten material shows a stratified temperature profile according to the different densities resulting from the different temperatures. In this regime the heat transfer is governed by the thermal conduction of the generated heat to the boundaries. In the upper part of the molten material a configuration similar to the heat transfer processes in the sun's and earth's mantles can be found. The hot material is cooled from the top which leads to lower densities at the bottom of the upper regime and a higher density close to the upper cooled boundary. This leads to a mixing of this upper region due to convective currents resulting in a turbulent mixed flow regime with an almost uniform temperature distribution.

During the last decades the heat transfer processes in molten core material have been of interest in nuclear research in an effort to improve safety measures in nuclear facilities [9], [13]. Numerous experimental studies were performed to understand the governing processes and to predict the behaviour of externally cooled molten core material. In addition numerical investigations adressing these experiments were performed and new models have been developed based on the experimental and numerical findings [14], [15].

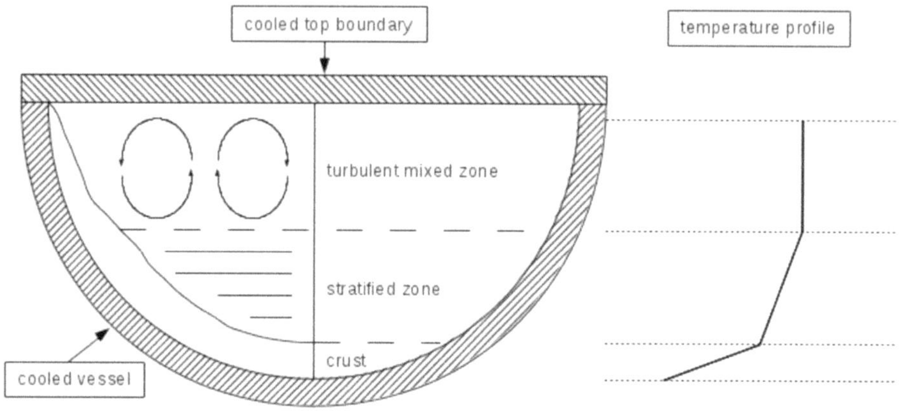

Figure 1.1: Different temperature and flow regimes in an internally heated fluid cooled through the adjacent boundaries

In additon to the research on the behaviour of molten core material, other concepts to improve the safety measures in *NPP*s in case of severe accidents have been investigated. Therefore, a plurality of numerical and experimental research studies has been performed. However, full scale severe accidents in *NPP*s, similar to crash tests performed in the automotive industry, cannot be performed. One reason is obviously the enourmous monetary cost of such an experiment, much more importantly, however, such tests would entail a tremendous and unjustifiable risk of causing radioactive contamination. Therefore, experimental test facilities focus on the investigation of specific phenomena in a model of the actual component [9]. For example, in the *LIVE*-facility the behaviour of molten core material is investigated in a 1:5 scaled model of the lower plenum of a *PWR*. The core material is simulated by a mixture of liquid salt, which is heated from the inside [16].

In order to predict the progression of an entire severe accident, system codes for *NPP*s have been developed. These system codes often use integral models derived from experiments or analytical and numerical investigations. Using these system codes an integral analysis of a severe accident can be performed [9]. The *MELCOR* (Methods for Estimation of Leakages and Consequences of Releases) [17] and ASTEC [18] codes are examples of such system codes. In comparison to the experimental or detailed numerical analysis of a specific phenomenon, these system codes show a lower accuracy and deficiencies at describing individual pheonomena. Szabó [19] showed for example that the specialized code GASFLOW provides a more precise distribution of the hydrogen in the containment during a severe accident compared to the results received when using the system code *MELCOR*. However, such specialized codes cannot simulate the entire *NPP*.

New computational developments and extended resources allow for the calculation of more complex geometries with detailed codes. However, the entire calculation of a *NPP* under severe accident conditions is not realisable with reasonable computational resources and calculation time since very long transients have to be considered. Even the simulation of the thermohydraulics within the core region with a detailed Computational Fluid Dynamics (*CFD*)-code is not realisable for the analysis of severe accidents. Therefore, simplified models have been developed to obtain more precise results than the system codes in a similar amount of computational time [20]. For the description of the heat transfer in molten core material a detailed analysis with a *CFD*-code is feasible but the computational effort is too high to use these in current safety analyses. However, based on the analysis of molten core material with *CFD*-codes and specific experiments, such as the *LIVE*-facility, more detailed models compared to those used in system codes can be developed. For example, models based on an effective diffusivity approach [21] or an effective convectivity approach [15] have been developed to model the flow phenomena in molten material based on empirical correlations. Such improved models can provide more accurate predictions of specific phenomena which might occur in a severe accident or even allow the simulation of additional phenomena, which cannot be taken into account by current system codes. Due to the extended computational resources the calculation time of these models is nowadays similar to the models used in the current system codes.

An approach to improve the current system codes is the coupling with other codes. At the Karlsruhe Institute of Technology (*KIT*) for example in order to improve the prediction of the hydrogen distribution in the containment the system code *MELCOR* was coupled with the *CFD*-code GASFLOW [19]. In this coupling *MELCOR* calculates the primary and secondary circuit and GASFLOW simulates the pheonomena in the containment building. The two codes are coupled at the leak in the primary circuit.

The development and coupling of new models to the system code *MELCOR* to improve its capabilities to simulate a severe accident in a *NPP* is the objective of this work. In coordination with the experiments performed in the *LIVE*-facility a new model to simulate the behaviour of molten core material in the lower plenum of the *RPV* is coupled to *MELCOR* and extended to simulate additional melt configurations.

The structure of this thesis is as follows: In the second chapter an overview of the state of the art in nuclear safety research will be introduced. The basic philosophy of safety in *NPP*s will be presented. A special focus will be on the *IVR*-strategy, which involves the cooling of a degraded core in the lower plenum of the *RPV* to stop the accident progression by preserving the integrity of the lower head structure. Selected experimental programs to investigate molten core material in the lower plenum will be summarized in a short overview. The *LIVE*-facility, which shall be used to validate the developed model as well as the coupled calculations, will be presented in detail. An overview of the results of the theoretical, numerical and experimental analysis of molten core material in the lower plenum will follow. The Phase-Change Effective Convectivity Model (*PECM*) developed at the Royal Institute of Technology (*KTH*) [15] - an example of a detailed model for the analysis of the behaviour of molten corium - will be explained in detail. The discussion of system codes for the simulation of severe accidents in *NPP*s with a special focus on the integral code *MELCOR* will be followed by the presentation of a methodology to characterize different coupling systems. At the end of this chapter an overview of the coupling interface in the code *MELCOR* will be given.

The third chapter will cover *MELCOR* simulations of different experiments carried out in the *LIVE*-facility. These calculations will show the necessity of an additional heat transfer model to take into account the convective heat transfer at the outer lower head surface to the adjacent liquid coolant. Furthermore, these simulations will show the deficiencies of this system code to predict the behaviour of molten material in the lower plenum in detail. This, however, is essential to evaluate the applicability of the *IVR*-strategy. Standalone simulations performed with the *PECM* will show that this model can provide the desired details. Hence, the objective of this work includes the expansion of the *MELCOR*-code by new models and the further development of these models.

The *LIVE*-L7v experiment will in the focus of the fourth chapter. Simulations performed on the experiment *LIVE*-L7v will demonstrate that the *PECM* shows significant deviations compared to the experimental data for the simulation of molten core material, which is cooled through the lower head and also from the top. A new approach taking into account the stratified region in the lower part of the molten pool willbe introduced. The new Stratified Approach for the Phase-Change Effective Convectivity Model (*S-PECM*) shall be validated by simulating the *LIVE*-L7v test. The results will be compared compared to the experimental findings and the data calculated by the original *PECM*.

In the fifth chapter the developed coupling system base on the tool *DINAMO* enabling the *PECM*-simulation of the molten core material within a *MELCOR* calculation will be presented. The coupled system will be tested by simulation of the experiment *LIVE*-L1. This experiment shall also be investigated by standalone calculations with *MELCOR* and the *PECM* in OpenFOAM. The validation database for the *PECM*, *MELCOR* as well as the coupled system will be expanded by the further simulation of the *LIVE*-tests L10 and L11.

The sixth chapter will demonstrate the applicability of the coupled *MELCOR-PECM* system to an entire *NPP*-simulation. Therefore an input deck for a generic *PWR* will be developed. In comparison with the standalone *MELCOR* approach a significantly different and more detailed behaviour of the molten material and the lower head will be observed.

Finally in chapter seven the results will be summarized and conclusions will be drawn.

2 Fundamentals and State of the Art

The present chapter will review the fundamentals related to this work. Furthermore the current state of research concerning severe accidents and the simulation of molten pool behaviour in the lower plenum shall be presented. A brief overview of severe accidents in a Nuclear Power Plant (*NPP*) and the relevant safety concepts will be presented. A more detailed discussion of the accident management strategy In-Vessel Melt Retention (*IVR*) willbe presented. This includes a description of the mechanism of this strategy as well as examples of commercial *NPP*s, which use the *IVR*. A variety of research has been conducted to investigate the *IVR*. Therefore, an overview of the experiments adressing the behaviour of the molten core material in the lower plenum will be presented. The experimental *LIVE*-facility at the Karlsruhe Institute of Technology (*KIT*), which was used to validate the simulations within this work will be discussed in detail. Based on the experimental work several models and correlations to describe the behaviour and heat transfer in a pool of molten core material have been developed. A selection of these correlations will be followed by a detailed discussion of the Phase-Change Effective Convectivity Model (*PECM*). Subsequently the integral method to simulate a severe accident in a *NPP* shall be discussed with special focus on the integral code *MELCOR*. At the end of this chapter the classification of the coupling compared to other code-to-code coupling methods will be presented and currently available couplings of integral codes and detailed simulation codes will be shown.

2.1 Nuclear Safety

The thermodynamic cycle used for electric energy generation in a *NPP* is similar to other thermal power plants. The electric turbine is powered by steam produced in the nuclear reactor. However, in *NPP*s the generation of steam by the fission of radioactive materials requires special safety measures. The radioactive materials have to be kept inside the facility to prevent contamination of the environment. Also, unlike other thermal power plants, in *NPP*s the heat source, which generates the steam, cannot be shut down entirely. Even after shutdown the reactor still produces a fraction of its operational thermal power due to radioactive decay of the core components. This decay heat can potentially evaporate the water in the primary circuit and even melt down the core if sufficient coolant is not provided [22].

The safety concept in *NPP*s is the so called Defense-in-Depth-approach. This approach includes the use of successive barriers to prevent the release of radioactive material from the facility. The objectives of this strategy have been defined as follows [23]:

- compensate potential human or component failures
- maintain the effectiveness of the barriers by averting damage to the plant and the barriers
- protect the public and environment in case the safety barriers are not fully effective

A schematic illustration of the barriers in the Defense-in-Depth-approach can be found in figure 2.1. The first barrier to retain the radioactive material is the fuel pellet itself. As long as the structure of the pellet matrix is intact no radioactive material can be released. The pellets in a fuel rod are surrounded by a cladding to hold back the radioactive material in case of failure of the pellet structure. The next barrier is the boundary of the primary circuit, which mainly coincides with the Reactor Pressure Vessel (*RPV*). The last two barriers, which can prevent the release of radioactive material to the environment, are the two containment vessels the so called containment system. These and more detailed information on the barriers as well as on the Defense-in-Depth-approach can be found in [24], [23] and [9].

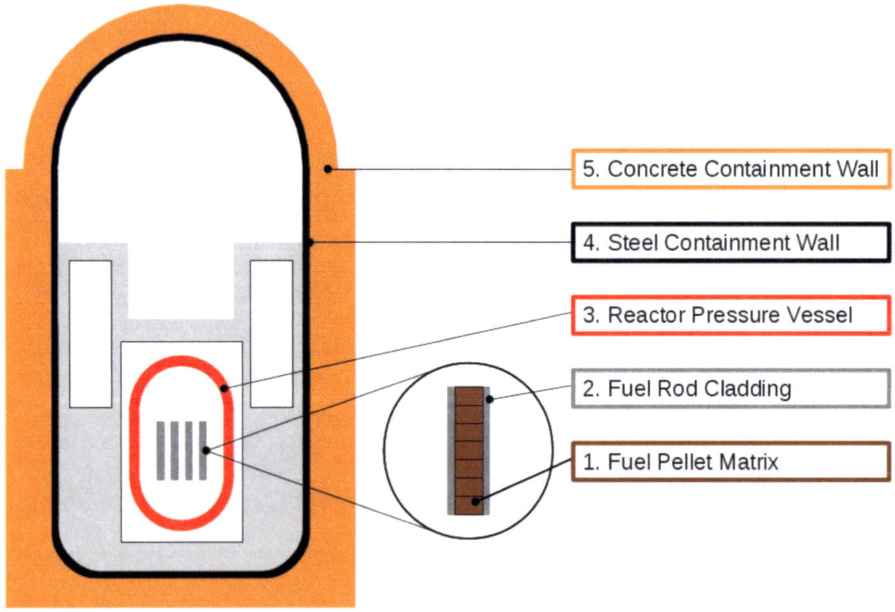

Figure 2.1: Successive barriers in the Defence-in-Depth approach according to [25]

The probability of occurrence of events and the resulting effects on the *NPP* are used to categorise hypothetical accidents. Design basis accidents may occure with a probability higher than 1 % within the lifetime of the *NPP*. The effects have to be addressed in the design phase of the *NPP* and a radiological impact outside of the power plant has to be excluded. Beyond design basis accidents are unlikely to occur (less than 1 % chance to occure during the life timeof the plant). These accidents may lead to core degradation, which is then classified as a severe accident [24]. Such a severe accident can lead to a failure of multiple safety barriers up to the release of a significant amount of radioactive material to the environment, if no sufficient coolant supply can be maintained [26].

Severe accidents are very unlikely to occur, however, accidents like TMI-2 (1979) [8], Chernobyl (1986) [9] or Fukushima (2011) [10] showed that they cannot be excluded entirely and that these accidents can have significant effects on the environment for decades. Therefore, despite the low probability, severe accidents have to be taken into account in the design and operational phase of a *NPP* ([27] and [28]).

In order to prevent, control and mitigate the effects caused by a severe accident Severe Accident Management (*SAM*) strategies have been developed. According to [29] the main objectives of *SAM* are:

- Prevent damage to the core
- Termination of core damage
- Maintaining the integrity of the containment as long as possible
- Minimizing on-site and off-site releases and their adverse consequences
- Achieving a long term stable state [10]

This *SAM* strategy has to be specified individually for each *NPP*, however some basic guidelines are common for most of the plants. These basic *SAM* actions are listed below according to Sehgal et al. [9].

- **Cooling a degraded core**

  The injection of water into the primary circuit is implemented in all water cooled *NPP*s as part of the *SAM* strategy. However, additional water can result in a higher hydrogen production due to the strongly exothermal reaction of zirconium-oxide with steam. Yet, sufficient water injected into the *RPV* can cool down the degraded core, stop further degradation and prevent the failure of the *RPV*. In order to inject the water into the primary circuit redundant passive systems or pumps at different pressure levels are installed. The pressure in the *RPV* can be decreased by secondary or primary side depressurization to enable the injection of the coolant. Another method to cool a degraded core is the flooding of the reactor cavity in order to cool the degraded core in the lower plenum from the outside. This *IVR* strategy is discussed in detail in the following chapter.

- **Management of combustible gases**

  This *SAM* strategy aims to reduce the amount of combustible gases, such as hydrogen H_2 or carbon monoxide CO, in the containment in order to prevent ignitions or explosions. In most of the western *NPP*s so called passive catalytic recombiners are installed to decrease the amount of combustible gases in the containment atmosphere. Other plants use igniters to induce combustion prior to a critical concentration of the gas or offer a sufficient containment size to prevent critical concentrations.

- **Management of containment temperature, pressure and integrity**

  In order to prevent an increase of the temperature and pressure in the containment and to preserve the integrity of the last barrier in many plants containment spray systems are used. These systems induce the condensation of steam in the containment. For long term cooling of the containment these spray systems can also be used as heat exchangers to avoid pressurization. Another *SAM* strategy used in the containment are fan coolers, which extract the heat to avoid high pressures due to non-condensable gases caused by Molten Core - Concrete Interaction (*MCCI*). In European plants containment venting is applied to manage these non-condensable gases.

- **Management of radioactive releases**

  To prevent the failure of the containment, it might be necessary to reduce the pressure by releasing some of the containment atmosphere to the environment. The amount of radioactive isotopes in the atmosphere can be reduced by spray systems. However, these spray systems are often designed for the early phase of a severe accident, for example to mitigate the effects of a Loss of Coolant Accident (*LOCA*). Another possibility to reduce the release of radioactive material to the environment are High Efficiency Particulate Air (*HEPA*) filters in the venting system.

The prevention of a severe accident in a *NPP* is part of current research and different phenomena are numerically and experimentally investigated. An overview of the classification of accidents in *NPP*s as well as basic mitigation strategies has been shown in this chapter. Further information on severe accidents and the different effects occurring can be found in the technical literature ([9], [30], [31] and [32]). In the next section the *IVR* strategy, which intends to stop the progression of a severe accident by external cooling of the *RPV* lower head, will be discussed in detail.

2.2 The In-Vessel Melt Retention Strategy

In the Three Mile Island Accident in Unit 2 (TMI-2) [8] about 45 % of the core melted and formed a melt pool in the core region. Part of the molten material relocated to the lower plenum. Investigations showed that the lower head of the *RPV* was able to retain the molten material and to prevent the progression of the severe accident into the containment. Due to this dramatic accident, methods to prevent or mitigate similar accidents in *NPP*s, were examined. One idea to stop the accident progression is the In-Vessel Melt Retention (*IVR*) strategy ([12] and [33]).

The objective of the *IVR* strategy is to prevent the failure of the lower head of the *RPV* during a severe accident in a *NPP*. In a hypothetical severe accident insufficient coolant supply leads to the melting of the core structures and the relocation to the lower plenum. In such an event, the *IVR* strategy suggests the flooding of the reactor cavity thereby transferring the decay heat from the molten core material through the lower head to the water in the adjacent cavity. With the retention of the molten material in the *RPV*, ex-vessel phenomena such as steam explosions in the containment or interactions of the molten corium with the concrete of the reactor cavity (*MCCI*) can be avoided [34]. According to the United States Nuclear Regulatory Commission (*U.S.NRC*) [33] in the Westinghouse AP-600 the safety measures in the containment such as spray systems are not related to the safety of the *NPP*. Due to the implemenation of the *IVR* strategy the conainment systems do not have to be considered in the safety concept.

For the retention of molten core material in the *RPV* the integrity of the lower head has to be tested against the possibility of a meltdown of the structure and a failure. This can be induced by mechanical stress or thermal load on the lower head structure due to the high temperature in the molten material. In molten core material temperatures of approximately 3000 K can be observed. Mechanical stress on the structure can be caused by weight of the relocated material or a high pressure in the *RPV*. Pressure peaks can be caused by a steam explosion due to the interaction of water in the lower plenum with molten core material. A combination of the thermal and mechanical load on the structure may lead to a thinning of the wall due to partial melting of the steel wall which is then no longer able to withstand the mechanical stress [7].

In order to realize the *IVR* strategy in a *NPP*, according to Rempe et al. [33] special technical requirements have to be met.

1. **Fast supply of coolant in the cavity**

   In case of a severe accident a quick and sufficient coolant supply to flood of the reactor cavity with water has to be guaranteed.

2. **Investigation of heat transfer from the lower head**

   A fast heat transfer from the lower head outer surface to the water in the reactor cavity has to be realized. This is especially challenging in the hemispherical part of a lower head. The steam bubbles created by nucleate boiling at the hot *RPV* wall rise along the curved part of the vessel before they are released into the bulk water. The heat transfer has to be fast enough to prevent the melt down and failure of the lower head.

3. **Influence of lower head insulation**

   The influence of the lower head insulation has to be identified. This insulation can influence the water flow and increase or decrease the cooling performance of the *IVR*. The insulation structure also has to withstand the force of the water flooding the cavity in order to prevent blockages or reduce the coolant flow along the vessel.

4. **Influence of structures attached to the lower head**

   The influence of structures in the lower head such as penetrations for instrumentation or control rods on the water flow along the outer surface of the lower head has to be determined to ensure their structural integrity and their effect on the cooling.

For the *NPP* Loviisa in Finnland of type VVER-440 with a thermal power of 510 MW the *IVR* strategy was approved in 1996. Prior to the installation of the *IVR* concept calculations and experiments were performed to investigate the structural integrity of the lower head in case of a severe accident. In order to implement the new safety measure, modifications had to be made to the reactor design. These changes included passive in- and outlet valves to ensure a continuous coolant circuit. A mechanism to lower the thermal insulation of the lower head in case of core uncovery was installed. This widens the gap between the insulation and the lower head to improve the coolant flow. Furthermore, a reactor depressurization system was implemented to reduce the mechanical stress on the lower head caused by high pressure in the primary circuit. In the cavity a detection system for debris was installed to detect leakage from the failed *RPV* ([34] and [33]).

Besides the Loviisa *NPP*, *IVR* is also used in the passive reactor concepts of Westinghouse AP-600 and AP-1000. For the AP-600 with an electrical power of 600 MW similar solutions as in the Loviisa reactor have been applied.

For the AP-1000 with an electrical power of 1000 MW and a thermal power of 3400 MW the concept had to be revised. The higher decay heat and the higher mass of the reactor core result in a higher thermal and mechanical load on the lower head structure in the event of a severe accident with *IVR*. Therefore, the insulation system of the lower head was redesigned in order to improve the guidance of the coolant flow. Additionally, the water level for the *IVR* in the cavity was increased along with an increased coolant mass flow [33]. The *IVR* concept of the AP-1000 was certified by the NRC in 2005 [35]. In China currently a reactor based on the AP-1000 design is under construction.

For high power reactors with an electrical power of above 1300 MW the realization of the *IVR* concept is very difficult due to the high thermal and mechanical loads on the lower head. This was already observed in the certification process of the AP-1000. Furthermore, the Institut de Radioprotecion et de Sûreté Nucléaire (*IRSN*) recommends a dry cavity to avoid the risk of a steam explosion in the containment. Steam explosions are capable to destroy the integrity of the steel and concrete containment walls. Therefore, competing designs use core catchers or spreading areas to cool the molten core material after the *RPV* failure ([35] and [36]).

The heat flux from the molten material through the lower head to the coolant is a key parameter in studies dealing with *IVR*. In order to predict this heat transfer a detailed knowledge of the behaviour of the molten material in the lower plenum is needed. This molten core material can be divided into two parts. The bottom part is the oxidic melt. Here the molten uranium oxide as well as oxidized Zirconium or Steel from the supporting core structures is found. The top part of the melt is represented by a metallic layer. In this layer the molten and non-oxidized zirconium and steel structures can be found. The heat transfer in the oxidic and metallic layer is governed by natural convection. Due to the cooler *RPV*-wall and the metalic layer in comparison to the oxidic melt a crust is formed at these boundaries dividing the oxidic part in a liquid and solidified region. A schematic view of the oxidic and metallic layer in the lower plenum with external cooling is shown in figure 2.2.

Different experiments and numerical studies have been carried out to investigate the behaviour of the molten material. The following section shows a selection of experimental facilities focusing on the investigation of the oxidic phase and the heat transfer through the lower head.

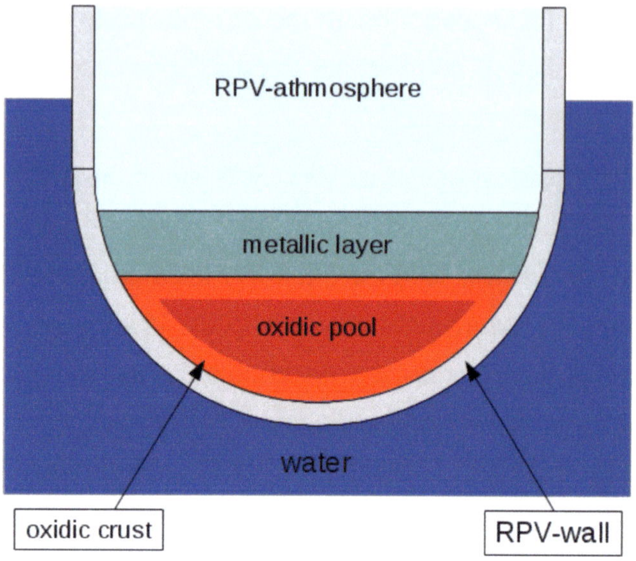

Figure 2.2: Material composition in the lower plenum during *IVR*

2.2.1 Experimental Facilities for In-Vessel Melt Retention

In this section experimental facilities investigating the melt pool behaviour in the lower plenum during the *IVR*-strategy for the molten core material are presented. Special focus will be put on the determination of the heat flux through the lower head.

- **ACOPO**

  In the ACOPO and mini-ACOPO test facility at the University of California the heat transfer in a 3-dimensional melt pool with natural convection is investigated. The ACOPO facility is a 1:2 scaled model of the lower plenum of a western type Pressurized Water Reactor (*PWR*) (AP-600) with a radius of 1 m. The cooling from the top and the vessel wall is uniform and no heating is applied to the simulant material [37]. As simulant material water is used in ACOPO whereas in the mini-ACOPO experiments also Freon-113 was used [38].

- **BALI**

  The BALI-facility located at the CEA in Grenoble is a 2-dimensional slice of a half hemispherical lower plenum with a radius of 2 m. In the experiments water, which is uniformly heated by direct current, is used to simulate the molten corium. The heated simulant can be cooled through the vessel walls and the top by an organic liquid to investigate the formation of an ice crust [39]. Another objective of the BALI experiments is to study the focusing effect caused by a metallic layer on top of molten oxidic material [40].

- **COPO**

  The experiments performed within this program can be divided into two parts. The COPO-I facility is a 1:2 scaled slice of the lower plenum of the Loviisa power plant that was approved for *IVR* measures [41]. The uniformly heated simulant material (solution of H_2O and $ZnSO_4$) can be cooled from the sides and from the top to investigate the heat transfer in the pool caused by natural convection [41]. The COPO-II experiments include two facilities. COPO-II-Lo is similar to COPO-I whereas COPO-II-AP is a 1:2 scaled 2-dimensional slice of the lower plenum of a western *PWR* (AP-600). In COPO-II the experimental facility is cooled by liquid nitrogen allowing the investigation of stratification in the melt pool and crust formation [42]. Experiments with immiscible layers were performed also in COPO-II [40].

- **EC-FOREVER**

  At the EC-FOREVER (European Commission' failure of reactor vessel retention) experimental facility of the Royal Institute of Technology (*KTH*) the failure of the *RPV* by molten core material in the lower plenum is investigated. As simulant a salt melt consisting of Calcium-Oxide (CaO) and Boron-Oxide (B_2O_3) is used. For cooling purposes the *RPV* can be flooded with water to cool the molten material from the top in order to provide similar cooling conditions as in the TMI-2 accident [43].

- **MIT CHF-flow loop**

  At the Massachusetts Institute of Technology (*MIT*) the Critical Heat Flux (*CHF*) along the outer surface of the lower head is investigated. Therefore, different lower head materials as well as different inclinations of the test section to represent specific sections of the lower head can be simulated. Besides the *CHF* the impact of the heat transfer on the lower head can also be investigated [44].

- **RASPLAV**

  The RASPLAV-200 test facility is a 2-dimensional slice of a hemisphere with a radius of 0.2 m. In this test facility, that can be cooled through the vessel walls and is insulated from the top, the heat transfer characteristics of different simulant materials are tested. As simulant material salt mixtures (e.g. $NaF - NABF_4$) or mixtures of uranium-oxide (UO_2), zirconium-oxide (ZrO_2) and zirconium (Zr) were used. With these materials the actual corium behaviour can be captured. The results of these small scale RASPLAV experiments were investigated afterwards in the larger RAPSLAV 1:2.5 test facility ([45] and [46]).

- **SIMECO**

  The experiments in the SIMECO-facility (Simulation of In-vessel Melt Coolability) at the *KTH* in Stockholm are performed to investigate the effects of boundary crusts and mushy layers on the heat transfer in a uniformly heated melt pool due to natural convection. Also the melt stratification is of special interest [47]. The facility is a 2-dimensional slice of a hemispherical lower plenum with a diameter of 62 cm and can be cooled from the sides as well as from the top to investigate the effect of a metal layer [48]. Besides the experiments with an eutectic and non-eutectic salt mixture of $NaNO_3$ and KNO_3 special tests were performed with up to 3 immiscible layers ([49] and [50]).

- **SULTAN**

  In the SULTAN test facility at the CEA in Grenoble the heat transfer from the lower head to the coolant is investigated. In the SULTAN experiments water under different conditions and with different mass flows is heated up in the test section while the heat transfer is monitored. With the SULTAN facility it is also possible to simulate different inclinations of the lower head [51].

2.2.2 The LIVE-Facility

The *LIVE* (Large Scale Experiments on In-Vessel Melt Relocation and Retention) facility at *KIT* was built to investigate the transient and steady state heat transfer in molten core material in the lower plenum of a *PWR RPV*. Another focus in the *LIVE* tests is the crust formation at the lower head wall and the effects of different relocation parameters ([52] and [53]). There are two experimental facilities in the *LIVE* program. The *LIVE*-3D test facility is a 1:5 scaled model of a western type *PWR* hemispherical lower head. *LIVE*-3D is the sole 3 dimensional model of the lower head, which can also heat the simulant. The *LIVE*-3D facility consists of two vessels. The inner vessel with a diameter of 1 m and a wall thickness of 0.025 m is made of stainless steel. The outer vessel serves as a cooling vessel to realize external cooling of the facility by water or air. The facility can be closed either by an insulation lid or a cooling lid to realize cooling from the top by water. A schematic view of the *LIVE*-3D test facility with the cooling lid can be found in figure 2.3.

Inside the test vessel a volumetric heating system is installed to simulate the decay heat. This heating system consists of 8 (formerly 6) layers of electrical heating coils. To realize an almost homogeneous power generation in the melt the distance between the heating layers and the

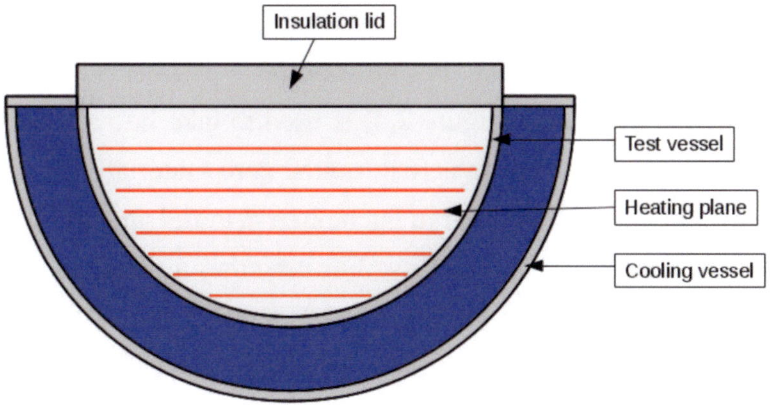

Figure 2.3: Schematic of *LIVE*-3D test vessel with insulation lid

windings is uniform at approximately 0.045 m. The heating planes can also be seen in figure 2.3. With this system a maximum heating power of 29 kW can be realized. For most of the experiments conducted in the *LIVE*-3D test facility a mixture of molten salt is used to simulate the molten core material. However in some experiments water was used as simulant material [54]. The salt mixture consists of sodium nitrate ($NaNO_3$) and potassium nitrate (KNO_3) [55]. Figure 2.4 shows the phase-diagram of this salt mixture. The *LIVE*-2D test facility is a slice with the same scale as *LIVE*-3D

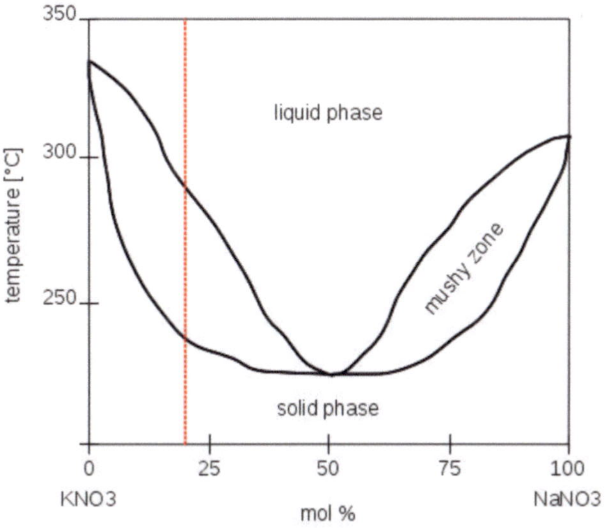

Figure 2.4: Phase-diagram of the KNO_3-$NaNO_3$ salt mixture

At the beginning of an experiment in the *LIVE*-3D facility the molten salt mixture is poured into the facility from an external furnace. For most of the experiments a salt mixture with 20% $NaNO_3$ and 80 % KNO_3 (red line in figure 2.4) is used to take into account the non-eutectic solidification process that is expected in actual molten core material. In order to monitor the temperature development and heat transfer in the molten material and the vessel approximately 100 thermocouples are installed in the facility. Furthermore, a measuring lance to detect the crust thickness as well as an optical and an infrared camera is installed. A more detailed description of the applied measurement systems can be found in [56].

Several experiments have been conducted in the *LIVE*-3D test facility. In *LIVE*-I1, at the beginning of the experimental program, water as simulant material was heated up inside the vessel. The hot water inside was cooled through the vessel wall by external water ([57]). *LIVE*-L1 and *LIVE*-L3 were conducted to investigate the effects of different pouring positions of the molten salt mixture ([58]). Different pouring amounts and multiple pourings on the transient behaviour of the molten material were examined in the *LIVE*-L2, L4 and L5 tests ([59] and [60]). A copper plate to vertically separate the material was used to investigate the effects of immiscible layers under different heating conditions in *LIVE*-L6 ([14]). The *LIVE*-L7 tests examined the influence of the top cooling lid on the melt behaviour using molten salt and water as simulant materials ([61]). Remelting of debris in the lower plenum by decay heat and relocation of additional molten material to the lower plenum was investigated in *LIVE*-L8 tests ([62]). *LIVE*-L9 showed the effect of a different salt mixture. Here, the eutectic salt composition of 50 % $NaNO_3$ and 50 % KNO_3 was used [63]. In the experiments *LIVE*-L10 and L11 the effect of sub-cooled water and boiling water to cool the facility through the lower head was investigated [64].

2.2.3 Correlations and Models for In-Vessel Melt Retention

The importance to capture the behaviour of the molten core material in the lower plenum to predict the integrity of the *RPV* during *IVR* has been shown in the previous chapter. In order to predict the progression of an actual severe accident with the *IVR* strategy applied, numerical models describing the behaviour of the molten material and the heat transfer from the molten material to the *RPV* wall are necessary. The heat transfer in the lower plenum is governed by natural convection. This convection is caused by the external cooling at the outer surface of the lower plenum and the internal heating of the melt caused by the decay heat. Natural convection phenomena can be described by the Grashof number (Gr) and the Prandtl number (Pr). The Grashof number represents the ratio of the bouyancy to the viscous forces in the material. The Prandtl number is defined as the relation between the viscous and the thermal diffusion rate.

Due to the decay heat the Dammköhler number (Da) is also needed to describe the molten material. This number is defined as the ratio of the internal heat generation to the heat transfer in the material. These numbers are expressed as ([65], [33]):

$$\begin{aligned} Gr &= \frac{g\beta(T_{max}-T_i)H^3}{\nu^2} \\ Pr &= \frac{\nu}{\alpha} \\ Da &= \frac{\dot{Q}H^2}{k(T_{max}-T_i)} \end{aligned} \tag{2.1}$$

g is the gravitational acceleration. The material parameters β, ν, α and k are the molten core material's thermal expansion coefficient, kinematic viscosity, thermal diffusivity and thermal conductivity respectively. H is a characteristic length of the investigated geometry. $\dot{Q}$ symbolizes the volumetrically applied decay heat of the molten material. In this case the vertical height of the molten pool is used. The temperatures T_{max} and T_i are the maximum temperature in the molten material and the temperature at the boundary of the pool at the inner suface of the lower head, respectively. The combination of the Grashof number and Prandtl number can be extended by the Dammköhler number resulting in the modified Rayleigh number (Ra). This modified Rayleigh number allows the classification of a corium pool and is defined as:

$$\begin{aligned} Ra &= Gr \cdot Pr = \frac{g\beta(T_{max}-T_i)H^3}{\alpha\nu} \\ Ra' &= Gr \cdot Pr \cdot Da = \frac{g\beta\dot{Q}H^5}{\alpha\nu k} \end{aligned} \tag{2.2}$$

In general the heat transfer can be characterised by the Nusselt (Nu) number, which is defined by [7]:

$$Nu = \frac{qH}{k(T_{max}-T_i)} \tag{2.3}$$

q is the average heat flux over the boundary of the investigated system. According to Rempe and Theofanous ([33] and [7]) the overall heat transfer from the molten corium with internal heating by the decay heat to the cooled walls can be characterised as a function of the Grashof number, Prandtl number and Dammköhler number. In a first approach the Nusselt number in an internally heated volume can be described as a function of the modified Rayleigh number.

$$Nu = f(Gr, Pr, Da) = C_A Ra'^{C_B} \tag{2.4}$$

C_A and C_B are empirically determined constants. Based on these fundamental assumptions several experiments and numerical studies have been performed to determine these constants. Some of these correlations are exemplarily shown in the folloing.

The first investigations on the heat transfer of an internally heated fluid, which is cooled externally, were carried out on a fluid layer to understand the fundamental phenomena governing the heat transfer. One of these experiments was carried out by Kulacki and Goldstein in 1972 [66], who investigated a fluid layer which was internally heated and cooled by the limiting walls. From this experiment a correlation for the heat transfer to the top and bottom boundary could be derived. For Prandtl numbers between 5.76 and 6.09 and modified Rayleigh numbers between $4 \cdot 10^4$ and 10^7 the correlation is defined by:

$$Nu_{up} = 0.436 Ra'^{0.228} \tag{2.5}$$

for the heat transfer to the upper boundary (Nu_{up})
and by

$$Nu_{dn} = 1.503 Ra'^{0.095} \tag{2.6}$$

for the heat transfer to the bottom cooled wall (Nu_{dn}).

Mayinger et al. [65] found a similar correlation in their experiments on the heat transfer in an internally heated fluid layer for a Prandtl number of 7 and a modified Rayleigh number between $8 \cdot 10^4$ and 10^{11}. The equations for the downward and upward heat transfer were determined as follows:

$$\begin{aligned} Nu_{up} &= 0.405 Ra'^{0.233} \\ Nu_{dn} &= 1.484 Ra'^{0.095} \end{aligned} \tag{2.7}$$

Further investigations were carried out in a rectangular fluid volume. The advantage of the rectangular setup is the possibility to investigate the heat transfer to the top and the bottom boundary of the volume. In addition, the heat transfer to the side walls can be investigated. Bachmann et al [67] investigated the heat transfer in a volume which was cooled through all boundaries (top, bottom and sides).

According to Bachmann et al. [67] the flow in a rectanglular geometry can be divided into three general regimes:

1. **The upper mixed regime**

   This regime in the upper part of the rectangular geometry is governed by turbulent structures and Rayleigh-Taylor instabilities, leading to a very good mixing of the fluid layer. The average temperature can be assumed as constant.

2. **The stratified regime**

   In the lower part of the rectangular geometry a stratification of the temperature can be found. However, this stratification is not entirely based on thermal conduction because of the high gradient.

3. **Flow at the side walls**

   The cooled fluid at the side walls induces a flow towards the bottom wall. This leads to a lower temperature in the regime close to the wall.

Due to these 3 regimes the heat transfer to the sides, the bottom and the top has to be described separately. Steinberner and Reineke [68] performed experiments on the heat transfer with a modified Rayleigh number in the range of $5 \cdot 10^{12} < Ra' < 3 \cdot 10^{13}$ and a Prandtl number of 7. For cooled walls at the side, top and bottom of the rectangular cavity the following correlations for the heat transfer could be derived:

$$\begin{aligned} Nu_{up} &= 0.345\, Ra'^{0.233} \\ Nu_{dn} &= 1.389\, Ra'^{0.095} \\ Nu_{sd} &= 0.85\, Ra'^{0.19} \end{aligned} \tag{2.8}$$

Experiments with the same setup, in which merely the side walls were cooled, lead to the same correlation for the sideward heat flux (Nu_{sd}):

$$Nu_{sd} = 0.85\, Ra'^{0.19} \tag{2.9}$$

For a hemispherical pool of liquid material that is heated internally and cooled through the side walls as well as the top boundary an illustration of the temperature distribution is reported [67]. The temperature in the semi-circular geometry shows the same three regimes identified in the rectangular geometry.

Mayinger et al. [65] performed experiments in such a semicircular cavity with cooled side and top walls and obtained the following heat transfer correlations for modified Rayleigh numbers of $10^7 < Ra' < 5 \cdot 10^{10}$ and a Prandtl number of 7:

$$\begin{aligned} Nu_{up} &= 0.36 Ra'^{0.23} \\ Nu_{dn} &= 0.54 Ra'^{0.18} \end{aligned} \tag{2.10}$$

Bonnet and Seiler [39] also investigated the heat flux in the semi-circular BALI-facility (chapter 2.2.1). In combination with simulations performed using the experimental data as boundary conditions they also identified the different regimes in the temperature distribution in the molten material. An illustration of the temperature regimes in a semi-circular lower head based on the work of Bonnet and Seiler [39] is shown in figure 1.1 on page 4.

Based on the measured data and data obtained from the ACOPO experiments Bonnet and Seiler [39] developed correlations for the 2- and 3-dimensional heat transfer to the vessel wall. The experiments were carried out with a Prandtl number of 0.2 and within a range of $10^{13} < Ra' < 10^{17}$ for the modified Rayleigh number.

$$\begin{aligned} Nu_{up} &= 0.383 Ra'^{0.233} \\ Nu_{dn} &= 0.116 Ra'^{0.25} \left(\frac{H}{R}\right)^{0.32} \quad \text{2-D geometry} \\ Nu_{dn} &= 0.131 Ra'^{0.25} \left(\frac{H}{R}\right)^{0.19} \quad \text{3-D geometry} \end{aligned} \tag{2.11}$$

In this correlation H is the height of the pool, whereas R symbolizes the circular radius of the pool calculated by the pools volume.

In a hemispherical cavity with an adiabatic boundary condition applied at the top wall and cooling condition at the curved walls Asfia et al. [69] determined the downward heat flux correlation for a Prandt number of 8 and modified Rayleigh number in the range of $10^{11} < Ra' < 10^{14}$ as:

$$Nu_{dn} = 0.54 Ra'^{0.2} \tag{2.12}$$

For a Prandtl number of 3 and a modified Rayleigh number of $2 \cdot 10^{10} < Ra' < 2 \cdot 10^{11}$, Gabor et al. [70] also investigated an internally heated fluid in a hemispherical geometry. Applying a free surface boundary condition at the top and cooling through the curved walls they determined the downward heat transfer correlation as:

$$Nu_{dn} = 0.55 Ra'^{0.15} \left(\frac{H}{R}\right)^{1.2} \tag{2.13}$$

The correlation derived by Jahn and Reineke (equation 2.10), Asfia et al. (equation 2.12) and Gabor et al. (equation 2.13) are very close. However, they are restricted to a specific Prandtl number. The dependence of the heat transfer on the Prandtl number was investigated at several institutions. Dropkin and Somerscales [71] performed experiments with different fluids (water, oil and mercury) and observed a noticeable effect of the Prandtl number on the heat transfer in the fluid. For a fluid layer heated from below and cooled from the top the following correlation for the upwards heat transfer was determined as:

$$Nu_{up} = 0.069 Ra'^{0.333} Pr^{0.074} \tag{2.14}$$

According to Dropkin and Somerscales [71] equation 2.14 is valid for a modified Rayleigh number between $3 \cdot 10^5$ and $7 \cdot 10^9$ and Prandtl numbers in the range between 0.02 and 8750. However, the inclination angle of the fluid layer affects the constants in equation 2.14. Based on these investigations Churchill and Chu [72] determined a correlation for the sidewards heat transfer by free convection to a vertical plate for arbitrary Prandtl numbers and a modified Rayleigh number of up to 10^{12}:

$$Nu_{sd} = \left(0.825 + \frac{0.387 Ra'^{\frac{1}{6}}}{\left(1 + \left(\frac{0.492}{Pr}\right)^{\frac{9}{16}}\right)^{\frac{8}{27}}} \right)^2 \tag{2.15}$$

The presented equations 2.5 to 2.15 are correlations to determine the entire heat transfer in the fluid to a specific boundary. In most system codes such as *MELCOR* [73] these correlations are used to determine the heat transfer from the molten corium to the adjacent boundaries. However, in order to predict the applicability of the *IVR* the heat transfer profile along the the lower head of the *RPV* is of interest. This is addressed in the work of Chawla and Chan [74], who developed a correlation for the local sideward heat flux based on the works of Eckert and Jackson [75].

$$Nu_{sd}^{local} = 0.508\,Pr^{\frac{1}{4}}\left(\frac{20}{21}+Pr\right)^{-\frac{1}{4}} Ra_y^{\frac{1}{4}} \tag{2.16}$$

The local Rayleigh number Ra_y is determined according to Chawla and Chan using the local Grashof number Gr_y:

$$\begin{aligned} Gr_y &= \frac{g\beta(T_{max}-T_B)H_{local}^3}{\nu^2} \\ Ra_y = Gr_y\cdot Pr &= \frac{g\beta(T_{max}-T_B)H_{local}^3}{\nu\alpha} \end{aligned} \tag{2.17}$$

The height H_{local} is the distance from the current position in the pool to the top boundary. The temperatures T_{max} and T_B define the maximum temperature of the pool and the temperature at the boundary, which is usually the solidus temperature of the molten material.

In addition to the heat transfer, the crust formed at the cooled surfaces has a significant influence on the heat transfer. In the following section the solidification process is discussed.

Solidification

The external cooling of the outer lower head wall also reduces the temperature of the adjiacent molten material. If sufficient cooling is provided, this can lead to a solidification of the melt in this region and the formation of a crust. According to the composition of the molten material different solidification mechanisms can be observed. In the LIVE-facility a salt mixture is representing the molten core material in the experiments. The phase diagram of this mixture is shown in figure 2.4. In this mixture the two relevant solidification mechanisms can be found.

- **Eutectic solidification process**

  At a percentage of about 50 % of $NaNO_3$ the liquid and solid phase are directly connected in figure 2.4. In this mixture the liquid material solidifies when the interface temperature between the liquid and solid regime is reached. This direct phase-change from liquid to solid is the eutectic of the mixture. This solidification behaviour is similar to the behaviour of pure materials, which can also be seen in figure 2.4 at 0 % and 100 % of $NaNO_3$.

- **Non-eutectic solidification process**

  Apart from the eutectic material composition in figure 2.4 a mushy zone between the solid and the liquid phase can be observed. In this case the liquid mixture is not directly solidified. The solidification process takes place between the solidus (lower boundary of the mushy-zone regime) and the liquidus temperature (upper boundary of the mushy-zone regime). If the temperature is between the solidus and liquidus temperature solid as well as liquid material can be found.

The solidification process influences the heat transfer to the cooled boundaries because the material properties in the crust differ from the molten part. The information presented in this section are based on [76] and [77].

A variety of experiments has been carried out to capture the heat transfer from an internally heated volume to the cooled boundaries. Also, several correlations have been derived from the experimental findings and theoretical analysis. These correlations are used in different models. One of these models, the *PECM*, will be discussed in the following chapter.

2.2.4 The Phase-Change Effective Convectivity Model

The Phase-Change Effective Convectivity Model (*PECM*) was developed at the *KTH* as part of the dissertation of Chi Thanh Tran in 2009 [15]. The objective of this model is to simulate the temperature development of the molten core material in the lower plenum using a reasonable amount of computational time and resources. The idea of the *PECM* is to reduce the effort needed for a detailed analysis of the molten material by Computational Fluid Dynamics (*CFD*)-methods. Therefore, instead of employing the equation system used for a typical *CFD* analysis of a molten pool, such as the momentum equations, the continuity equation and the energy equation, just the energy equation shown in equation 2.18 has to be solved[78].

$$\underbrace{\frac{\partial(\rho c_p T)}{\partial t}}_{\text{transient behaviour}} + \underbrace{\frac{\partial(\rho c_p T U_x)}{\partial x} + \frac{\partial(\rho c_p T U_y)}{\partial y} + \frac{\partial(\rho c_p T U_z)}{\partial z}}_{\text{convection}} = \underbrace{\nabla(k\nabla T)}_{\text{diffusivity}} + \underbrace{Q_V}_{\text{heat source}} \tag{2.18}$$

In addition to the thermal conductivity k and the specific heat capacity c_p the velocities U_x, U_y and U_z in the convection terms are needed to solve the equation system on a numerical grid. Usually in *CFD*-simulations these velocities are determined by solving the continuity and momentum equations. In the *PECM* these velocities are approximated by empirical correlations. Cheung et al [79] already divided the molten pool in a stratified region, which is driven by the thermal diffusivity of the molten material, and a turbulent mixing region with an almost uniform temperature, which is governed by convection. This division is introduced in figure 1.1 on page 4.

To eliminate the velocities in the energy equation Cheung et al. [79] introduced an effective thermal diffusivity α_{eff}. The effective diffusivity is calculated according to the following correlation:

$$\begin{aligned} \alpha_{eff} &= \alpha \cdot Nu && \text{turbulent mixing region} \\ \alpha_{eff} &= \alpha && \text{stratified region} \end{aligned} \tag{2.19}$$

α is the thermal diffusivity of the molten material and Nu the Nusselt number calculated by an empirical equation. This model was implemented in the SCDAP code [80]. An empirical correlation by Kulacki and Nagle [81] is used to calculate the Nusselt number. The velocities in the conductivity terms are now represented by the effective diffusivity. Therefore, in the energy equation 2.18 the convective terms are not taken into account.

A similar approach was developed by Bui and Dinh [82] at the *KTH* prior to the development of the *PECM*. The Effective Convectivity Conductivity Model (*ECCM*) uses the effective diffusivity for the sidewards heat transfer in the molten material to neglect the sideward velocity U_x. By multiplying the thermal conductivity of the fluid with an empirical local Nusselt number, the effective sideward thermal conductivity $k_{\text{x, eff}}$ is calculated:

$$k_{\text{x, eff}} = k_x \cdot Nu_{sd}^{local} \tag{2.20}$$

The local Nussel number Nu_{sd}^{local} is determined according to the correlation by Chawla and Chan (2.16). The upward and downward heat transfer is approximated by characteristic velocities. These characteristic velocities U_{up} and U_{dn} for the upward and downward heat transfer can be calculated as ([83] and [78]):

$$\begin{aligned} U_{up} &= \frac{\alpha}{L}(C - Nu_{up}) \\ U_{dn} &= \frac{\alpha}{L}\left(\frac{CNu_{dn}}{Nu_{up}+Nu_{dn}} - Nu_{dn}\right) \end{aligned} \tag{2.21}$$

This equation uses the thermal diffusion α, the height of the molten pool L, the upward and downward Nusselt numbers Nu_{up} and Nu_{dn} as well as the constant model parameter C. For the upward heat transfer a correlation by Kulacki and Emera [84] was used whereas for the downward Nusselt number the correlation by Steinberner and Reinke (ref. equation 2.8) was implemented. This model is used in the research code MVITA [85].

At the Helmholtz Zentrum Dresden-Rossendorf (*HZDR*) the *ECCM* was modified and implemented into the ANSYS code to perform coupled thermal and mechanical simulations in the molten material and the lower plenum by Willschütz et al. [21]. In contrast to the *ECCM* the upward and downward heat transfer was also simulated by an effective diffusivity. In the stratified region no additional diffusivity was added in vertical direction, whereas the diffusivity facing the side wall was determined by the correlation 2.20 in the *ECCM*. In the turbulent mixed region the vertical diffusivity was multiplied by the upward Nusselt number Nu_{up}. Willschütz et al. used correlations by Bernaz [86] for the calculation of the upward and sideward Nusselt number.

In the *PECM* the convective heat transfer in the molten material is also approximated by characteristic velocities. Unlike the methodology in the *ECCM*, the *PECM* does not use effective diffusivities to calculate the heat transfer. The overall heat transfer in the *PECM* is governed by correlation 2.8 by Steinberner and Reineke [78]. In order to calculate the characteristic velocities for the *PECM* the pool of molten material has to be divided into an upper (H_{up}) and lower (H_{dn}) part (see figure 1.1). The total height of the pool (H_{pool}) and the calculated Nusselt numbers [85] are used to determine the height of these pool regions.

$$\begin{aligned} H_{up} &= \frac{H_{pool} \cdot Nu_{up}}{Nu_{up} + Nu_{dn} + Nu_{sd}} \\ H_{dn} &= H_{pool} - H_{up} \end{aligned} \tag{2.22}$$

The characteristic velocities for the heat transfer in the molten pool can be calculated using this height and the width of the pool W_{pool} as:

$$\begin{aligned} U_{up} &= \frac{\alpha}{H_{pool}} \left(Nu_{up} - \frac{H_{pool}}{H_{up}} \right) \\ U_{sd} &= \frac{\alpha}{H_{pool}} \left(Nu_{sd} - \frac{2H_{pool}}{W_{pool}} \right) \\ U_{dn} &= \frac{\alpha}{H_{pool}} \left(Nu_{dn} - \frac{H_{pool}}{H_{dn}} \right) \end{aligned} \tag{2.23}$$

In equation 2.23 the calculation of the sideward characteristic velocity U_{sd} correlation 2.16 was used to calculate the Nusselt number taking into account the heat transfer distribution along the curved lower head surface. Due to the fact that the characteristic velocities do not represent the actual flow in the molten material, the heat transfer by convection is implemented in the *PECM* as an additional source term to the energy equation 2.18. The energy added to the system by the convective heat transfer is subtracted uniformly from the system as a heat sink to prevent an additional heat source [87]. The energy equation solved on the *PECM* is written as:

$$\underbrace{\frac{\partial(\rho c_p T)}{\partial t}}_{\text{transient behaviour}} = \underbrace{\nabla(k\nabla T)}_{\text{diffusivity}} + \underbrace{Q_{DCH}}_{\text{decay heat}} + \underbrace{Q_{CON}}_{\text{local energy of convection}} - \underbrace{Q_{ADD}}_{\text{average energy of convection}} \tag{2.24}$$

This energy equation is similar to the energy equation of a solid material. This emphasises that in the *PECM* the difference between the liquid and the solid phase of the molten material is determined by the additional source term in the energy equation. Therefore, in the *PECM* merely equation 2.24 has to be solved, which leads to very low calculation times compared to traditional *CFD*-calculations. The calculation times were speeded up by as much as a factor of 200 [15]. However, for simulations with the *PECM* a numerical grid of the calculation domain remains necessary.

In the *PECM* the phase-change in non-eutectic material compositions (ref. figure 2.4) can also be taken into account. Dependent on the local temperature of the melt a factor determines the fraction of liquid material in the cell [85]. This fraction f_L is calculated according to equation 2.25 and can be seen in figure 2.5.

$$f_L = \frac{T - T_{solid}}{T_{liquid} - T_{solid}} \quad \text{for } T_{solid} < T < T_{liquid} \tag{2.25}$$

T_{solid} and T_{liquid} is the solidus and the liquidus temperature of the specific material composition, respectively.

With the *PECM* it is possible to locally resolve the energy equation. However, a finer nodalisation of the calculation domain compared to lumped parameter codes such as *MELCOR* is necessary. Due to the reduction of the equation system to one equation, the calculation time is significantly reduced compared to *CFD*-calculations. At the *KTH* the linear equation system of the *PECM* is implemented as a User Defined Function (*UDF*) into the commercial *CFD*-program ANSYS FLUENT. The *PECM* has been validated by the simulation of the

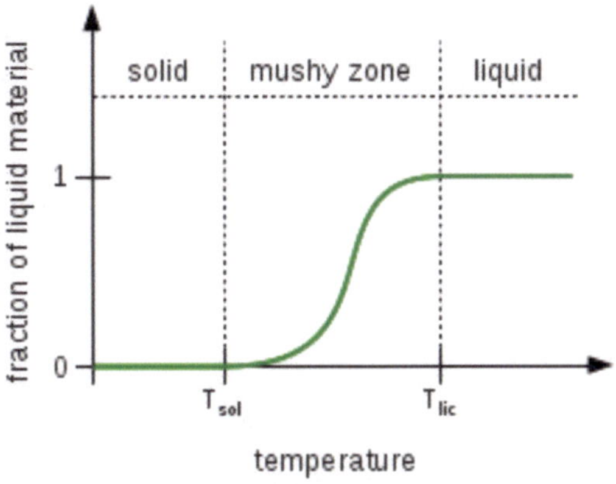

Figure 2.5: Fraction of liquid material depending on the local temperature

SIMECO experiments and the experiment *LIVE*-L1 [87]. In addition to simulating the oxidic pool the *PECM* has the possibility to take into account the metallic layer on top of it [88].

2.3 Simulation of a Severe Accident

This chapter adresses methods to simulate a severe accident in a *NPP*. The programs used in such simulations can be divided into two basic categories: Mechanistic and parametric codes [24].

Mechanistic codes use best estimate models to simulate specific phenomena within a severe accident as accurate as possible. These programs allow the calculation of the detailed progression of the severe accident within the boundaries of the code. Parametric code systems use phenomenological parametric models to simulate the integral behaviour of the entire *NPP* [89].

In Probabilistic Safety Analyses (PSA), parametric codes are used to calculate the overall progression of the severe accident based on different initial and boundary conditions. Therefore, parametric programs have to ensure short calculation times to handle a variety of different input parameters in an adequate amount of computational time. On the other hand mechanistic codes are used within PSA to perform detailed simulations of selected scenarios based on the previous integral simulations with a parametric code. In this chapter integral codes (parametric codes) will be discussed in more detail especially the integral code *MELCOR*, which was used in this work.

2.3.1 Integral Codes

The analysis of severe accidents in nuclear facilities with simulation programs is an important tool to determine the safety of the facility and investigate the risk of radioactive material leaking into the environment. Due to the complexity and size of nuclear facilities it is currently not possible to simulate an entire *NPP* with a high resolution in a reasonable amount of time. Therefore, lumped parameter codes are often used to analyse hypothetical and actual severe accident in nuclear facilities. Compared to detailed simulation tools such as *CFD*-programs these codes use a very coarse nodalisation of the nuclear facility. The connections between the Control Volumes (*CV*s) are usually calculated by 1D-equations [90]. On the one hand, this coarse nodalisation does not allow an exact resolution of the occurring phenomena. On the other hand, it allows the simulation of the overall trends in associated complex systems such as the primary circuit, the secondary circuit and the containment [9]. For example, the entire containment of a *NPP* is usually nodalised by approximately 30 nodes, whereas a *CFD*-simulation of a containment with ANSYS CFX4 uses about 160 000 cells ([91] and [92]).

Due to the coarse nodalisation and the simplified models in most of the integral codes the simulated time is often longer than the calculation time. This is important for parameter studies within Probabilistic Safety Analyses of Stage 2 (*PSA-2*). Within such an analysis, numerous calculations are performed to take into account different accident progressions and determine their risk of occurrence. Analyses of this type can also be used to determine the efficiency of the applied safety measures to control a severe accident. Furthermore, the low calculation time of integral codes can be used to perform parameter studies on the influence of model or boundary parameters ([18], [9] and [19]). Using integral codes, general information about the progression of an accident can be obtained. The results of previous calculations with an integral code can provide boundary conditions for detailed simulations to verify the validity of the used models.

There are different types of integral codes. Codes such as ATHLET-CD, COCOSYS or RELAP merely treat specific parts of the accident progression. For example, ATHLET-CD simulates the primary and secondary circuit, whereas COCOSYS was developed to simulate the phenomena occurring in the containment. A coupling between ATHLET-CD and COCOSYS is developed at the Gesellschaft für Anlagen- und Reaktorsicherheit (*GRS*) in order to simulate an entire accident progression in a *NPP* [90].

Other codes such as *MELCOR* [93], MAAP [94] or ASTEC [18] were developed to capture the entire accident progression. However, due to the plurality of models in the codes used to capture the different occurring phenomena in a severe accident these codes show a significant dependence on the input parameters [9].

2.3.2 The severe Accident Code MELCOR

The program *MELCOR* (Methods for Estimation of Leakages and Consequences of Releases) is developed and provided by Sandia National Laboratories (*SNL*). *MELCOR* is one of the leading codes for the integral analysis of severe accidents in a *NPP*. It is widely used for *PSA-2* in nuclear facilities ([93] and [73]). The development of *MELCOR* started in 1982 [17] as a reaction to the severe accident in the Three Mile Island power plant in block 2 ([8] and [11]). The first version of the program was released in 1989. The current version *MELCOR* 2.1 was released in 2009. This version is the first which is entirely written in the programming language FORTRAN-95. ([9]). The next version of *MELCOR* 3.0 is currently under development [95].

In *MELCOR* a variety of models to capture the progression of a severe accident and to analyse accident mitigation methods are implemented. However, most of these models are quite simple in order to ensure fast simulation results. Due to insecurities in the simplified models and the coarse nodalisation, *MELCOR* simulations are very sensitive regarding changes in the model parameters or the *MELCOR* system time step [19]. An example of the influence of the system timestep is presented in chapter B. *MELCOR* is divided into different packages, which each simulate specific phenomena in a severe accident [96]. At the system time step these packages of *MELCOR* are coupled explicitly. Between two system time steps each package can pursue its own time step advancement.

During the *MELCOR* development not only new models have been implemented into the system code. Also, simplified versions of entire stand-alone codes, such as the program CONTAIN, which takes into account the thermal hydraulic phenomena in the containment, were introduced into the *MELCOR* source code. A diagram showing the development of *MELCOR* can be found in figure 2.6.

The main advantage of *MELCOR* compared to other system codes such as MAAP [94] is the possibility that *MELCOR* can simulate an entire severe accident in a nuclear reactor. This means simulations of the primary circuit, the secondary circuit as well as the containment are possible in one calculation. Most of the other integral codes are often limited to the primary and secondary circuit (ATHLETH-CD [97]) or to the containment (COCOSYS [98]).

At the Institute of Nuclear and Energy Technologies (*IKET*) the version 1.8.6 of the *MELCOR* code, released in 2006 [9], is used. The code of this version is still partly written in FORTRAN-77 ([95]). The advantage of this version is the possibility to couple *MELCOR* to other codes. The coupling interface was already available in *MELCOR* version 1.8.5 ([99]). However, in *MELCOR* 1.8.6 the coupling interface had to be reactivated by Szabó [19]. At the IKET the source code of *MELCOR* 1.8.6 is available, which is necessary for the further development of the coupling interface.

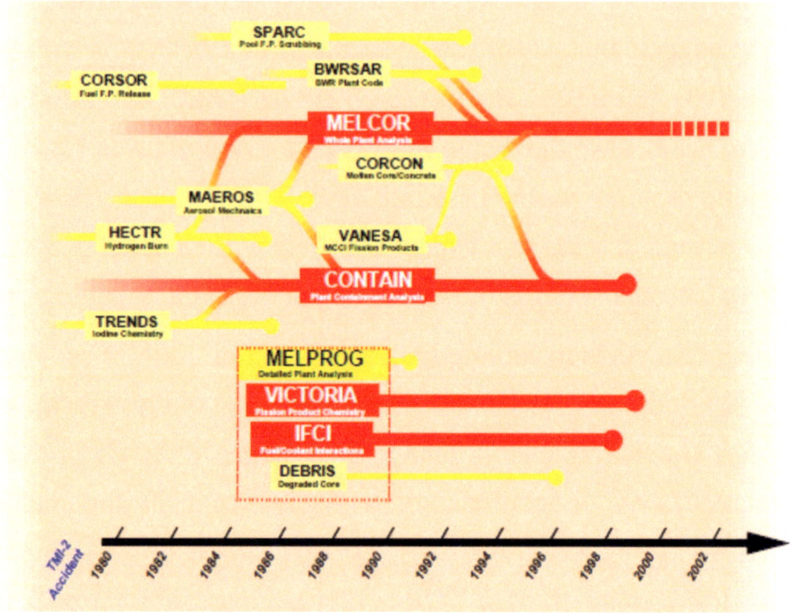

Figure 2.6: Diagram of the *MELCOR* development [17]

An important package of *MELCOR* is the Control Function (*CF*) module. The *CF*s can be processed by all packages of the code and are used amongst others to define complex safety systems in a nuclear reactor. The pumps which ensure a specific water level in the steam generator during reactor operation are controlled by this package, example.

In order to couple *MELCOR* to other programs, *CF*s are used to define the coupled variables and to send and receive the current coupling parameters [19]. Within this work a method was developed to define new *CF*s in *MELCOR* in order to perform a more detailed evaluation of the simulations and to expand the amount of parameters that can be transferred via the coupling interface [100].

2.4 Code to Code Coupling

In nuclear research different models are coupled to investigate the physical processes in *NPP*s. The system codes, such als MELCOR, as described in the previous chapter, consist of different interacting models which are coupled inside the program. Since the beginning of the development of such system codes multiple models have been coupled to describe the transient behaviour of the *NPP* [101]. The coupling of new models or programs to system codes is still of great importance in current nuclear research. In the european THINS-project an entire work package is dedicated to the improvement of system codes by code to code coupling [102].

As with most technical processes, different physical pheomena influence the behaviour of the technical system in *NPP*s as well. In order to numerically represent the entire system in detail all of these phenomena have to be considered in the simulation model. For example in order to simulate the behaviour of the human heart the fluid dynamics of the blood as well as the movement of the heart muscle have to be taken into account. At the *KIT* the *Karlsruhe Heart Model* (KaHMo) uses a *CFD*-code to capture the behaviour of the blood. The heart muscle is represented by a structural mechanic model. At the inner surface of the heart muscle the mechanical model is coupled to the CFD-code. This coupling also allows the simulation of the interaction of the structural and the fluid dynamics regions [103]. The interaction of fluid flow with solid structures is also of interest in the analysis of turbines or the thermal behaviour of aircraft components. In order to numerically simulate such systems also coupled codes are used ([104] and [105]).

Another application of coupled code systems is the reduction of the simulation time. Christian et al. [106] investigated seismic waves caused by fractures in materials. In order to reduce the calculation time a slow-running detailed code, which treats the regions with fractures, is coupled with a fast finite element code, which is used in the surrounding regions.

In order to simulate the phenomena, which can occur in nuclear reactors during operation or within severe accidents, several specific models have been developed. For the simulation of an entire *NPP*, however, it is necessary to simulate all the occurring effects and also to take into account the interactions between these phenomena. For example, the thermohydraulic phenomena in the core region directly affect the neutron kinetics which in return influence the behaviour of the thermohydraulics. Different coupled code systems have been developed to capture this interaction [107].

Furthermore, the prediction of hydrogen in the containment is another important topic in current nuclear reactor safety research. Coupled code systems are used to provide a more detailed description of the hydrogen behaviour ([108] and [19]).

The European project NURESIM [109], which started in 2005, is developing a *European Reference Simulation Platform for Nuclear Reactors*. In this project different codes for the analysis of *NPP*s are coupled to simulate various possible situations such as a hypothetical severe accident. For example in order to simulate the transient of a *NPP* up to three different code are coupled to receive a more detailed description compared to the standalone calculations. Currently additional coupling processes to other codes are developed to extend the amount of processes that can be simulated with this platform [110]. Therefore, especially in nuclear applications the code to code coupling is an important part of current research.

According to Avramova and Ivanov [111] the coupling of multiple codes is a very complex task requiring additional programming, computational and also mathematical effort. In the following chapters different ways to characterize a coupling process are presented. Afterwards the coupling interface in *MELCOR* is discussed in detail.

2.4.1 Coupling Characterisation

Coupled code systems can be realized in different ways. The common part of all couplings in nuclear engineering is the exchange of physical properties and calculation results as boundary conditions for the coupled code. These boundary conditions can improve the calculation results and the possibilities of the code compared to a stand-alone simulation [111]. In this chapter different strategies to couple two codes are presented.

The structure of the coupled system can be distinguished in three different categories [112]:

- **External Iterations**

  In this coupled system no exchange of data is performed during the simulation. The results of a stand-alone calculation in one code are used as boundary conditions in the coupled code. This is often done in the form of time-dependent tables. Iterations with updated boundary conditions are performed until the convergence criterion is reached. These simulations come along with a high computational effort and are very time consuming. However, no changes have to be made to the codes to realize the coupling.

- **Coupling of separate Codes**

  Unlike the external coupling methodology, in this coupled system the data exchange is performed during the simulation time. According to Grigic and Kozlowski [112] as well as Ivanov and Avramova [113] external programs are often used to control the coupling progress and coupling exchange. The codes in this coupled system perform stand-alone calculations with boundary conditions, which are updated at every coupling time step by the coupled code. This method is less time consuming than the external iterations strategy and the interaction between the different calculated phenomena can be addressed directly during the coupled simulation. However, changes have to be made to the source code of the program to realize the coupling. Also, an additional program to control the coupling exchange might have to be executed during the coupled simulation.

- **Integration**

  In this coupled system one code is directly implemented into the source code of the other one. In order to realize this, the source code of both programs has to be available and an extensive programming effort has to be made, as this requires significant changes to the structure of these codes. Due to the direct integration this coupling is very flexible to changes in the boundary conditions and merely one code has to be executed.

Another way to classify the coupling system is the use of the coupled programs. According to Ivanov and Avramova [113] a coupling can be external or internal. In the external coupling the calculation domain of the stand-alone programs is extended. This kind of coupling is used to couple a code, which simulates the phenomena in the primary and secondary circuit of a *NPP* for example with another code, which calculates the phenomena in the containment [19]. In this coupling approach the boundary conditions at the interface between the calculation domains of the two codes are exchanged. This coupling methodology requires minor changes to the source code of the participating programs, however, numerical instabilities can occur.

In the internal coupling the calculation domains overlap and the two codes are closely connected. An example of this coupling methodology is performed in this work by calculating the failure of the lower core support plate by a coupled model. In this case the thermal and mechanical stress calculated by *MELCOR* is used as a boundary condition. Based on these data the lower core support plate failure function is simulated and then sent back to *MELCOR* where the rest of the *NPP* is calculated. This coupling methodology is very direct and precise, however, a lot of data has to be exchanged between the codes and a higher programming effort is needed to establish the coupling [114].

The time progress of the coupled simulation can be executed in three different ways. There are explicit, semi-implicit and implicit couplings [113].

In the explicit coupling, the boundary conditions calculated by one code are used in the coupled codes for the following time steps. Between two coupling time steps the coupled boundary conditions are constant. The jump from one coupled value to the one calculated for the next time step can cause numerical instabilities. This can be reduced by using small coupling time steps.

In the implicit coupling, an equation system is established over the coupling interface. With this equation system the boundary conditions in the codes are calculated by both coupled programs. Using the implicit coupling methodology numerical instabilities due to sudden jumps in the values can be avoided. However, this method goes along with high computational costs and development effort.

The semi-implicit coupling strategy is a mixture of the explicit and implicit approach. With the equation system over the coupling interface, just boundary conditions which tend to cause numerical instabilities are calculated. The other boundary conditions participating in the coupling process are exchanged explicitly. With the semi-implicit methodology the computational and developing effort is reduced compared to the fully implicit method, however, the effort is higher than in the explicit method ([115], [113] and [116]).

The time step advancement has also to be taken into account when classifying a coupling process. The coupling exchange can either be synchronous or asynchronous [117]. In the asynchronous time step advancement a coupling time step is defined. All programs participating in the coupling have to meet this time step. Between two coupling time steps each program can pursue its own time step advancement. In the synchronous coupling methodology the time steps in all the coupled programs are the same. The synchronous coupling allows a better connection between the coupled codes due to the updates of the boundary conditions in every time step. However, the programming effort to realize a synchronous coupling is enormous [116].

According to Weaver [116] the calculations in the coupled codes can be performed parallel or sequential. During the parallel coupling the programs exchange boundary conditions at the coupling time step. Until the next coupling time step the simulations in the coupled programs are executed parallelly. In the sequential coupling the leading program sends the boundary conditions to the other codes. Then the calculations are performed in the coupled codes. At the end of the coupling time steps the boundary conditions are sent back to the leading code. This code then advances to the next coupling time step.

Often additional programs are used, in order to establish a coupling between two codes. These communication programs were developed, according to Szabó [19], to realize parallel computing on computer clusters. The advantage of such programs are predefined coupling routines and processes. In nuclear engineering the Parallel Virtual Machine (*PVM*)-protocol [118] or the Message-Passing Interface (*MPI*)-standard [119] are often used.

2.4.2 The coupling Interface in MELCOR

The coupling interface in *MELCOR* version 1.8.5 was developed to couple *MELCOR* to the system Code RELAP5-3D [120]. In this coupled system the *MELCOR* code was used to simulate the phenomena in the containment, whereas the RELAP5-3D code simulated the primary and secondary circuit of the *NPP* [99]. Therefore, the coupling routine in *MELCOR* 1.8.5 was developed according to the coupling interface in RELAP5-3D, which is based on the methodology presented by Weaver [121]. A communication program based on the

PVM-standard (PVMEXEC) [122] is used to establish the coupling. The coupling interface in RELAP5-3D is preferably used to couple codes for the simulation of the containment. Besides the coupling with *MELCOR* 1.8.5, RELAP5-3D was also coupled with the system code GOTHIC [123] and the *CFD*-codes ANSYS FLUENT [124] and ANSYS CFX [125].

The coupling between *MELCOR* 1.8.6 and RELAP5-3D developed by Cole [99] can be characterized, according to the previous chapter, as external, explicit and asynchronous. In *MELCOR* the variables exchanged via the coupling interface have to be defined as *CF*s. In order to receive data via the coupling interface a new type of *CF* was implemented in the *MELCOR* source code. The functionality of the coupling interface in *MELCOR* 1.8.5 was tested by the coupling of two separate *MELCOR* calculations and compared to stand-alone results. This test was also performed by Szabó [19] for the coupling interface in *MELCOR* 1.8.6. Besides the initial coupling of *MELCOR* with RELAP5-3D, Rodríguez [126] also performed simulations with this coupled system to verify its correct implementation and functionality.

The coupling exchange based on the *PVM*-standard was redesigned for the use of the *MPI*-standard by Young et al. [127]. This was done due to the advantages of the *MPI*-standard concerning parallel calculations on a single workstation. However the *MPI* based coupling interface still uses the methodology developed by Weaver [121] for the exchange of the coupling parameters. In order to control the execution of the coupled programs and to coordinate the coupling exchange the communication program MPIEXEC was used. This new coupling system was used to couple *MELCOR* with other codes developed by *SNL* such as GILA [128]. The coupling of two *MELCOR* instances was also carried out with the *MPI* based coupling interface [127].

After the update from *MELCOR* version 1.8.5 to version 1.8.6 the coupling routines developed by Young et al.[127] were still available in the code, however, due to missing variable declarations a coupling exchange could not be executed. This problem was addressed by Szabó [19], who analysed the *MELCOR* 1.8.6 source code and applied modifications to ensure a correct functionality of the coupling process. Furthermore, the amount of coupling parameters exchanged via the coupling interface was increased from 50 to 500 to allow for the coupling of complex programs. The coupling interface was then used by Szabó to couple *MELCOR* 1.8.6 to the *CFD*-code GASFLOW [129] to improve the prediction of the hydrogen concentration in the containment during a severe accident. In this coupling the primary and secondary circuit were calculated by *MELCOR* and the containment was simulated by GASFLOW. The coupling parameters of the external, explicit and asynchronous coupling were exchanged at the position of a leak in the primary circuit of the *NPP* [130]. More detailed information on the coupling interface in *MELCOR* and the coupling process itself are presented by Szabó [19].

3 Simulation of the LIVE-facility

In this chapter first the simulations performed with *MELCOR* version 1.8.6 on the *LIVE*-facility will be presented. Therefore, an input deck for the experimental facility will be developed. The input deck shall be tested by the simulation of the *LIVE*-I1 experiment. In this experiment water is heated inside the facility and cooled by externally provided water. These simulations will show the need for the modification of the heat transfer from the lower head to an adjacent liquid coolant. Afterwards the *LIVE*-L1 experiment will be simulated with *MELCOR*. The simulation results will show significant deviances to the experimental data. These findings will show that a new methodology to simulate the behaviour of molten material in the lower plenum during a severe accident in a Nuclear Power Plant (*NPP*) is necessary. The Phase-Change Effective Convectivity Model (*PECM*) has proven its capability to provide a more detailed description of the molten core material in the lower plenum. This model will therefore be used to recalculate the *LIVE*-L1 test.

3.1 The MELCOR input for the LIVE-facility

In order to perform simulations of the *LIVE*-facility in *MELCOR* an input deck has to be developed first. This model of the *LIVE*-facility includes the geometric and physical properties of the experimental facility. Figure 3.1 is an illustration of the thermohydraulic structure of this input deck.

According to the setup in the experiments the inside of the vessel has a direct connection to the environment, which is modeled by a time-independent Control Volume (*CV*), in order to ensure an almost constant atmospheric pressure during the entire experiment. The vessel of the *LIVE*-facility is also modelled by a single *CV* which can be filled up with water or air depending on the setup in the simulated experiment. On the outside of the vessel another *CV* for the coolant is located. This *CV* is connected to two additional time independent *CV*s. The velocity in the Flow Path (*FL*) between *reservoir 1* and the *coolant CV* is used to realise the coolant supply, which can either be air or water depending on the simulated experiment.

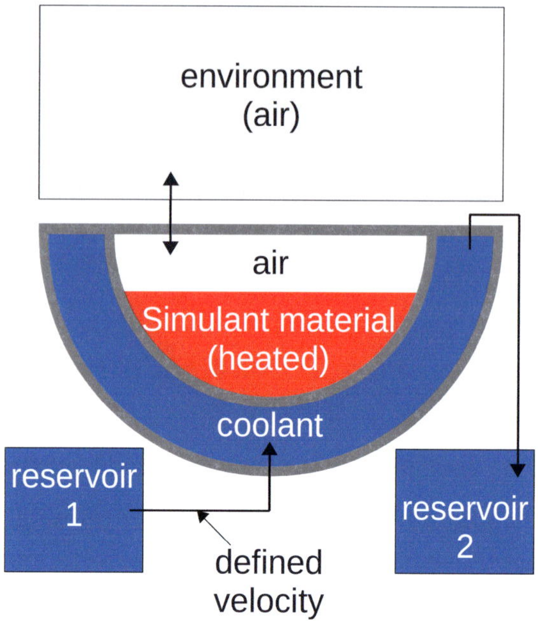

Figure 3.1: Thermohydraulic structure of the *MELCOR* input deck of the *LIVE*-facility

The inside of the *LIVE*-vessel is modelled in *MELCOR* by two overlapping packages. In the thermohydraulic package the behaviour of air and water in the vessel, as well as the communication with the environment, is simulated. The COR-package calculates the molten salt and the vessel wall. During a *MELCOR* simulation these packages internally exchange parameters. The nodalisation of the *LIVE*-vessel in these two packages is illustrated in figure 3.2.

In the COR-Package the *LIVE*-vessel is divided into 8 axial levels and 8 radial rings. An additional thin axial level is located on top of the *LIVE*-vessel to meet the requirements of *MELCOR*. The surrounding heat structure is made of steel and nodalised by 6 nodes. As already mentioned, the thermohydraulics are represented by a single *CV*. Simulations are performed with a more detailed nodalisation of the thermohydraulics package in the *LIVE*-vessel leading to numerical instabilities and unrealistic results [131].

For the computations of *LIVE*-experiments employing a salt mixture to simulate the molten corium, the input deck has to be modified. In *MELCOR* at the beginning of the simulation no molten pool or debris material can be located in the lower plenum. Therefore, an additional axial level outside the *LIVE*-vessel was introduced to provide the salt mixture in solid form at the start of the calculation. Due to the lack of a core support plate this material immediately drops into the *LIVE*-vessel and is then heated up by the heating power which can be introduced

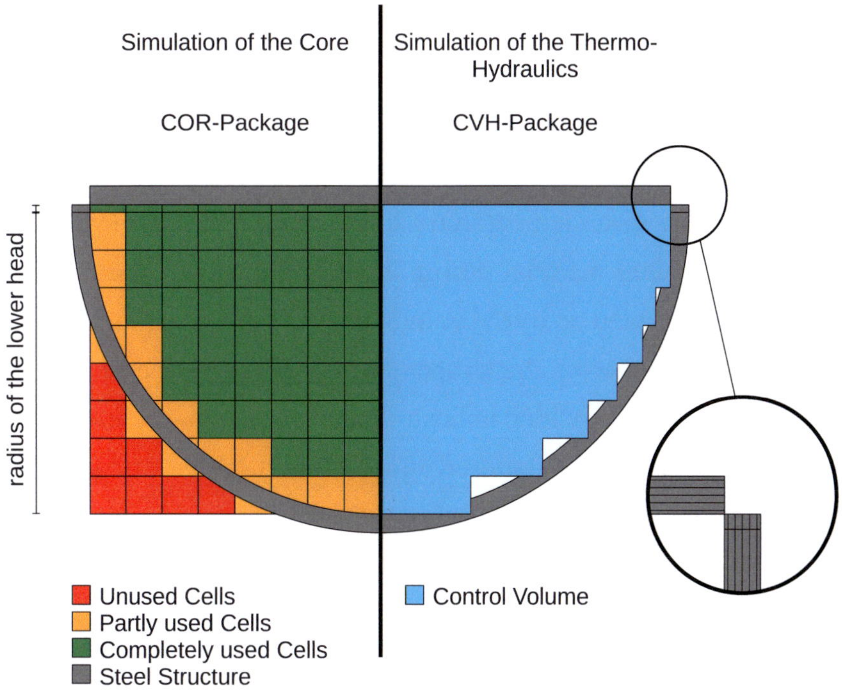

Figure 3.2: Nodalisation of the *LIVE*-vessel for simulating the core and the thermohydraulics

by the decay heat package or by a Control Function (*CF*). The properties of the salt mixture of $NaNO_3$ and KNO_3 are defined as a new fuel material in the *MELCOR* input deck. In the data base of the *LIVE*-experiments not all of the material properties of the salt are available. These material properties are obtained from the data base of the *LIVE*-facility from publications by Sehgal and Yang [132], Rodgers and Janz [133], Kenisarin [134], Jriri et al. [135] and Kawakami et al. [136].

3.2 Simulation of LIVE-I1 with MELCOR

The input deck for the *LIVE*-facility in *MELCOR*, as described in the previous chapter, is tested by simulating the experiment *LIVE*-I1. In this experiment water was used as a simulant material in the vessel of the *LIVE*-facility. This water with a pool height of 31 cm, which corresponds to a volume of 120 l, was heated up by the volumetric heating system with different heating powers. A steady state configuration could be found when a heating power of 10 kW was applied [57]. During the steady state external cooling was provided by water with a constant coolant supply of 10 cm^3 at approximately 23°C.

The simulations on the steady state of *LIVE*-I1 are performed with the *MELCOR* version 1.8.6. In figure 3.3 the temperature evolution of the water inside the *LIVE*-facility, the temperature of the coolant as well as the mean temperatures of the inner and outer surface of the *LIVE*-vessel in contact with the simulant material during the calculation are shown. The water inside the facility and the vessel is heated up by the volumetric heating power until the boiling temperature of the water is reached. However, the temperature of the coolant does not significantly change over the entire simulation time, which indicates that no heat transfer from the *LIVE*-vessel to the coolant is calculated. Due to the monitored temperature difference of about 77°C this simulated behaviour is physically unrealistic.

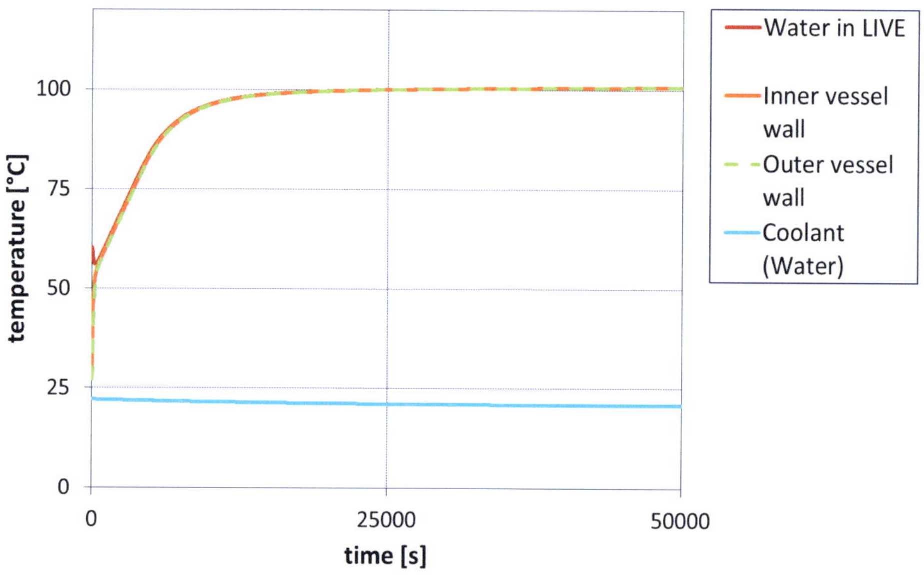

Figure 3.3: Temperature development in experiment *LIVE*-I1 simulated by *MELCOR*

The reason for this unsatisfactory and unrealistic behaviour of the simulation can be found in the “Reference-Manual“ of the *MELCOR* code [73]. The heat transfer between the outer surface of the Reactor Pressure Vessel (*RPV*) and the adjacent liquid coolant is calculated by a so called ”boiling“ model. This model merely predicts a heat transfer from the vessel to the coolant if the temperature of the outer vessel surface exceeds the saturation temperature of the coolant. In this case boiling of the liquid coolant can be observed. This model works well for the simulation of the heat transfer in actual *NPP*s, but in order to simulate the *LIVE*-facility this model cannot be applied. The heat transfer model in *MELCOR* 1.8.6 is therefore modified to take into account the heat transfer below the saturation temperature of the coolant which is observed in the most *LIVE*-tests. For the "sub-cooled" heat transfer, equation 3.1 was implemented into the *MELCOR* source code.

$$Q = \alpha \cdot A \cdot (T_V - T_C) \cdot \Delta t \tag{3.1}$$

The transferred heat Q is dependent on the heat transfer coefficient α, the surface area A and the current time step Δt. The driving force of the heat transfer is the difference between the temperature of the outer vessel surface T_V and the temperature of the coolant T_C. The heat transfer coefficient α is dependent on the properties and physical state of the involved materials. Therefore, an option is implemented in *MELCOR* to determine this heat transfer by a user defined *CF*. The 9th input parameter for the *CORLHDxx* section, which is currently not used in *MELCOR* 1.8.6, can now hold the negative number of a *CF* to specify the heat transfer for this segment. An example for the expanded input deck can be seen in figure 3.4.

```
*
*          *** Lower Head Structure ***
*
*          IS     TLH     RADLH   ICVCAV
CORLHD01   1      298.15  0.0621  300     0   0.0   0   0   -9999
CORLHD02   2      298.15  0.1241  300     0   0.0   0   0   -9999
CORLHD03   3      298.15  0.1862  300     0   0.0   0   0   -9999
CORLHD04   4      298.15  0.2066  300     0   0.0   0   0   -9999
CORLHD05   5      298.15  0.2483  300     0   0.0   0   0   -9999
CORLHD06   6      298.15  0.2921  300     0   0.0   0   0   -9999
CORLHD07   7      298.15  0.3104  300     0   0.0   0   0   -9999
CORLHD08   8      298.15  0.3507  300     0   0.0   0   0   -9999
CORLHD09   9      298.15  0.3725  300     0   0.0   0   0   -9999
CORLHD10   10     298.15  0.3945  300     0   0.0   0   0   -9999
CORLHD11   11     298.15  0.4281  300     0   0.0   0   0   -9999
CORLHD12   12     298.15  0.4345  300     0   0.0   0   0   -9999
CORLHD13   13     298.15  0.4538  300     0   0.0   0   0   -9999
CORLHD14   14     298.15  0.4861  300     0   0.0   0   0   -9999
CORLHD15   15     298.15  0.4966  300     0   0.0   0   0   -9999
*
```

Figure 3.4: Use of the new heat transfer model in the *MELCOR* input deck

In this example the heat transfer coefficient of the heat transfer is defined by *CF* no. 9999. The modified heat transfer for sub-cooled conditions is now used to recalculate the experiment *LIVE*-L1. The temperature development in this calculation can be found in figure 3.5.

In contrast to the temperature development found in traditional *MELCOR*, the modified heat transfer correlation allows *MELCOR* to simulate the heat transfer from the outer vessel wall to the adjacent coolant. This leads to a heat up of the the external water. Due to the external cooling, the water in the *LIVE*-facility does not reach its boiling temperature. This ,however, was not monitored in the *LIVE*-I1 experiment [57]. Compared to the findings in the *LIVE*-I1 test, the simulation overestimates the pool's temperature, which is shown in figure 3.6.

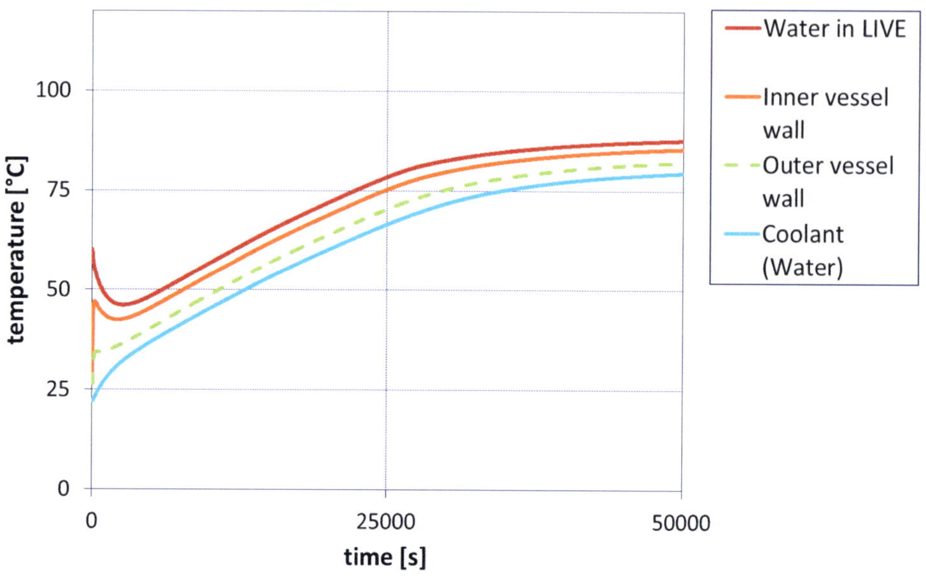

Figure 3.5: Temperature development in experiment *LIVE*-I1 simulated by *MELCOR* with modified heat transfer model

The simulations of the *LIVE*-I1 experiment with *MELCOR* show that it is not possible with *MELCOR* version 1.8.6 to predict the temperature development in the *LIVE*-facility. An extension of the *MELCOR* source code by introducing an additional heat transfer equation for sub-cooled heat transfer is necessary to capture the heat transfer from the outer *RPV* wall to the adjacent liquid coolant. With the modified *MELCOR* version the steady state observed in *LIVE*-I1 can be reproduced. However, a stratification of the temperature cannot be obtained due to the coarse nodalisation. Also, the temperature in *MELCOR* exceeds the temperature measured in the experiment.

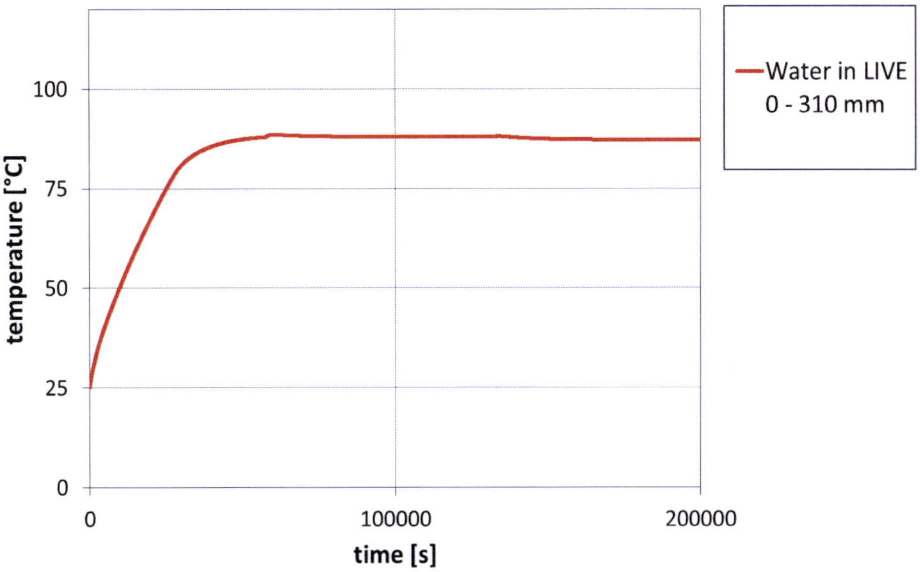

(a) Temperature in *MELCOR*

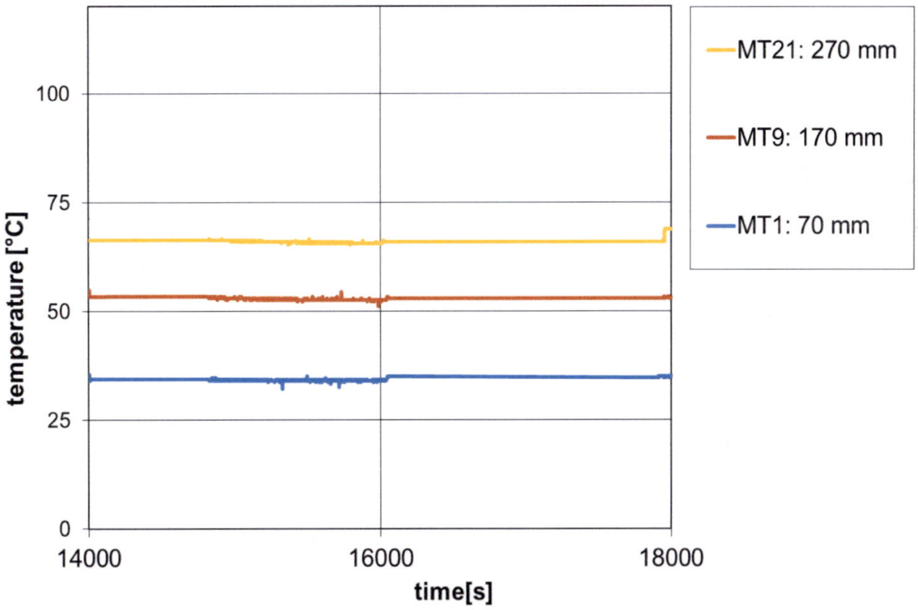

(b) Temperature at different elevations in *LIVE*-I1 [57]

Figure 3.6: Comparison of the temperatures in the steady state of *LIVE*-L1 with the *MELCOR* simulation

3.3 Simulation of LIVE-L1 with MELCOR

LIVE-L1 was the first experiment, which was carried out with salt as simulant material [54]. At the beginning of this test 120 l of molten salt was poured into the facility. This resulted in a height of the molten material in the *LIVE*-vessel of approximately 31 cm. The volumetric heating system heated the salt with different heating powers. From the outside the *LIVE*-vessel was cooled by water which was externally supplied with a rate of about 10 cm^3/s at 23°C. During the experiment the facility was closed by the insulation lid. According to the observations by Fluhrer et al. [58] a stable steady state was observed at heating powers of 10 and 7 kW. In this chapter the simulation result in *MELCOR* of the steady state with a heating power of 10 kW will be presented and compared to the experimental findings.

Figure 3.7 shows the temperature development in the molten salt calculated by the *MELCOR* version 1.8.6. According to the definition of the *MELCOR* input in chapter 3.1 at the start of the simulation solid salt drops into the lower plenum and is then heated up volumetrically. After the melting of the entire salt in the facility the external coolant region (ref. figure 3.1) is flooded with water and the constant coolant support is established. In figure 3.7 a steady state can be observed after about 20 000 s of simulation time. Due to the assumption of a uniform temperature in the entire molten material the local temperature distribution cannot be obtained in *MELCOR*

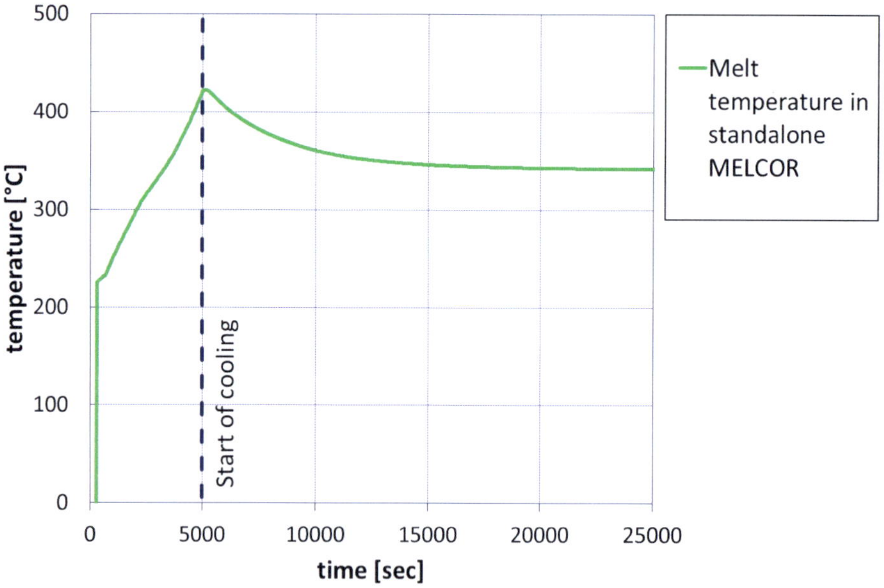

Figure 3.7: Temperature development in the molten pool for *LIVE*-L1 in *MELCOR*

The temperatures found in the *LIVE*-L1 test for the steady state with a heating power of 10 kW at different positions in the *LIVE*-vessel are shown in figure 3.8. Compared to the experiment the temperature calculated by *MELCOR* even overestimates the highest temperature measured in *LIVE*-L1.

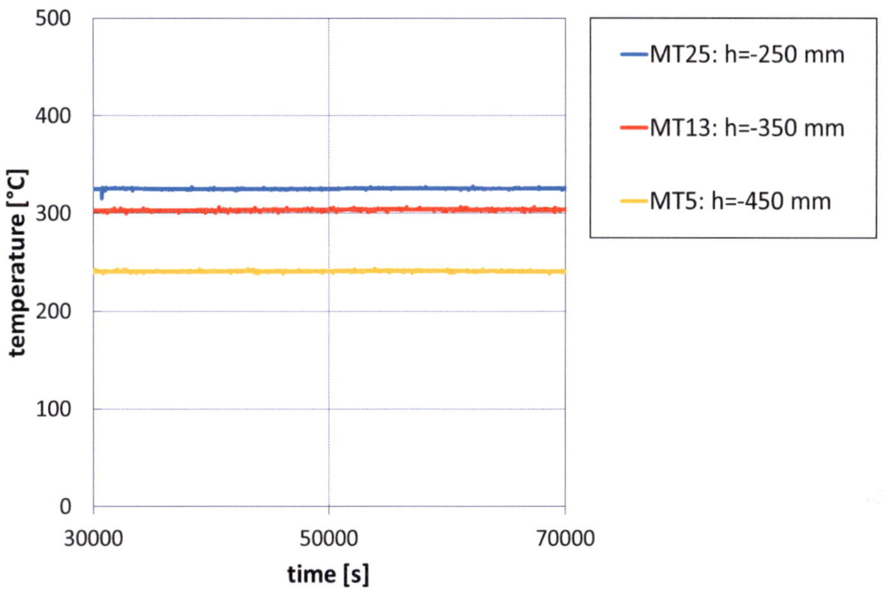

Figure 3.8: Temperatures of the molten salt in *LIVE*-L1 at different positions [58]

The simulation of the *LIVE*-L1 test shows that with *MELCOR* it is not possible to simulate the *LIVE*-test in detail. In *MELCOR* the molten core material is treated as one uniform continuum. Thus a uniform temperature distribution and uniform physical properties are determined in *MELCOR* for the corium. In *MELCOR* an eutectic material composition is assumed. Hence a fixed melting temperature is used. In an actual core melt however a non-eutectic solidification process similar to the salt mixture used in *LIVE*-L1 is assumed. The uniform temperature and the eutectic solidification process do not allow for the formation of a mushy zone between the crust at the cooled vessel wall and the fully molten material. Furthermore, despite the implementation of a crust formation methodology in *MELCOR* [73] no crust is formed in the simulation of the *LIVE*-L1 test. These are major deficiencies *MELCOR* shows when it comes to the detailed simulation of molten core material in the lower plenum of the *LIVE* facility. Therefore a different approach is required to evaluate the applicability of the In-Vessel Melt Retention (*IVR*) strategy. The objective of this work is to improve the modelling of molten core material in *MELCOR* by introducing new models to the system code and to further develop the existing models.

3.4 Simulation of LIVE-L1 with the PECM

The simulations of the *LIVE*-tests I1 and L1 showed that a stand-alone *MELCOR*-calculation cannot represent molten core material in sufficient detail. A detailed description of the molten material in the molten core is necessary, for instance, to investigate the applicability of the *IVR* strategy in a *NPP*. A detailed Computational Fluid Dynamics (*CFD*)-calculation is capable of providing the necessary information on the pool behaviour. However, due to the long calculation times of *CFD*-simulations such codes are not an option. The *PECM* (ref. chapter 2.2.4) can resolve the local temperatures in the molten core material in a reasonable amount of calculation time. Based on the temperature distribution in the molten pool the *PECM* allows for the observation of crust formation processes at the lower head surface. A model to treat the mushy zone between the crust and the fully molten material allows for the simulation of eutectic as well as non-eutectic solidification processes. This qualifies the *PECM* to be used instead of the *MELCOR*-model for the simulation of molten core material in the lower plenum of the *RPV* or the *LIVE*-vessel.

3.4.1 Implementation of the PECM in OpenFOAM

In the *PECM* the temperature profile is calculated on a numerical grid similar to the ones used in *CFD*-calculations. The nodalisation in *MELCOR* as presented for the *LIVE*-facility in chapter 3.1 cannot provide the required fine grid. Therefore, an external grid has to be used to solve the energy equation in the *PECM*. Furthermore, in the *PECM* a non-linear equation system for the temperature field has to be solved. This requires the use of a high-performance solver. At the Royal Institute of Technology (*KTH*) these challenges were met by implementing the *PECM* as a User Defined Function (*UDF*) in the *CFD*-code ANSYS FLUENT.

The intention of this work is to control the execution of the *PECM* in an external program and to modify the source code of the model. Therefore, an alternative to the the closed commercial *CFD*-software ANSYS FLUENT is preferred. The free available *CFD*-software OpenFOAM provides fast linear equation solvers and can also handle the numerical grid required for the simulations with the *PECM*. Additionally the open-source character of OpenFOAM allows direct modifications of the solver routines, which can be used for the implementation of the *PECM*. OpenFOAM further provides a tool to generate numerical meshes. The free post-processing software ParaView is part of an OpenFOAM installation.

A multi region solver is modified to implement the *PECM*-routines into OpenFOAM. This solver can take into account the molten material in the lower plenum as well as parts of the adjacent wall of the *RPV*. The source code of the multi region solver is modified to only solve the energy equation.

In order to use the *PECM* for modelling the convective heat transfer from the heated melt to the lower head a source term has to be implemented in the energy equation. The geometric dimensions of the molten pool are captured by a specific algorithm, which is developed in this work. In contrast to the approach in ANSYS FLUENT at the *KTH* it is also possible to capture non-connected pools in the lower plenum. The physical and geometric properties are used to calculate pool specific modified Rayleigh- and Nusselt-numbers for the heat transfer to the cooled boundaries. These numbers are used to calculate the characteristic velocities for the convective heat transfer. The heat sources induced by the convection are determined using the characteristic velocities (ref. chapter 2.2.4). Afterwards the total energy added by the convective heat source is uniformly subtracted from the entire pool.

To ensure the correct implementation of the *PECM* in OpenFOAM the same simulation setup was calculated with the modified solver in OpenFOAM and the *PECM* in ANSYS FLUENT. The test setup is based on the experiment *LIVE*-L1 [54]. At the beginning of the simulations 120 l of salt is located inside the facility with a temperature of 350°C. The upper boundary of the salt melt shows a fixed temperature of 280°C and the outer lower head surface cooled by a constant temperature of 80°C. The salt melt is heated with a power of 10 kW and a steady state is reached after 10 000 seconds. Both simulations in OpenFOAM and ANSYS FLUENT use the same numerical grid shown in figure 3.9. This grid was developed using the meshing software ICEM-CFD.

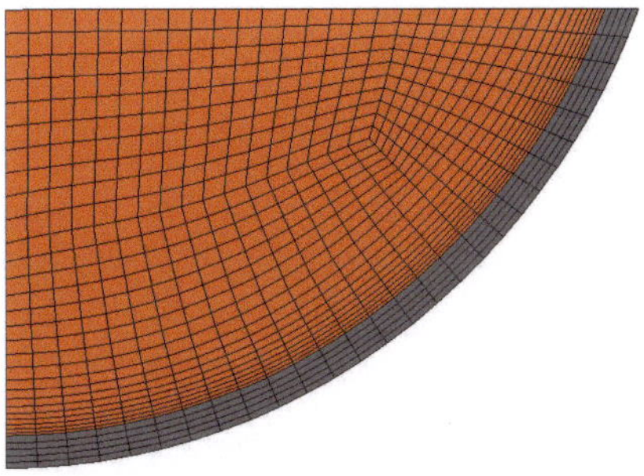

Figure 3.9: Grid for experiment *LIVE*-L1 used for the verification of the correct implementation of the *PECM* in OpenFOAM

To compare the simulation results different horizontal and vertical evaluation lines are placed in the calculation domain. The location of these lines to analyse the temperature profiles in the melt can be seen in figure 3.10.

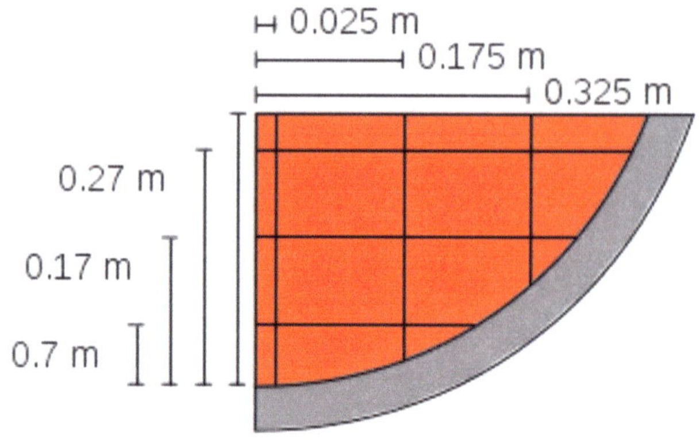

Figure 3.10: Positions of the evaluation lines to analyse the temperature profiles in the *PECM*

Figure 3.11 shows the vertical temperature profile in the molten salt along the vertical line at a distance 0.175 m from the symmetry axis of the calculation domain. In comparison to ANSYS FLUENT the temperature gradient towards the lower head is slightly overestimated in OpenFOAM. This leads to a higher temperature at the wall and a slightly thinner crust. Due to the higher temperature close to the wall the mean temperature in OpenFOAM is slightly lower than the mean temperature calculated by ANSYS FLUENT. Despite these deviances, which are most probably caused by the different discretisation and solving methodologies in the two *CFD*-codes, the temperature profiles are in exellent agreement.

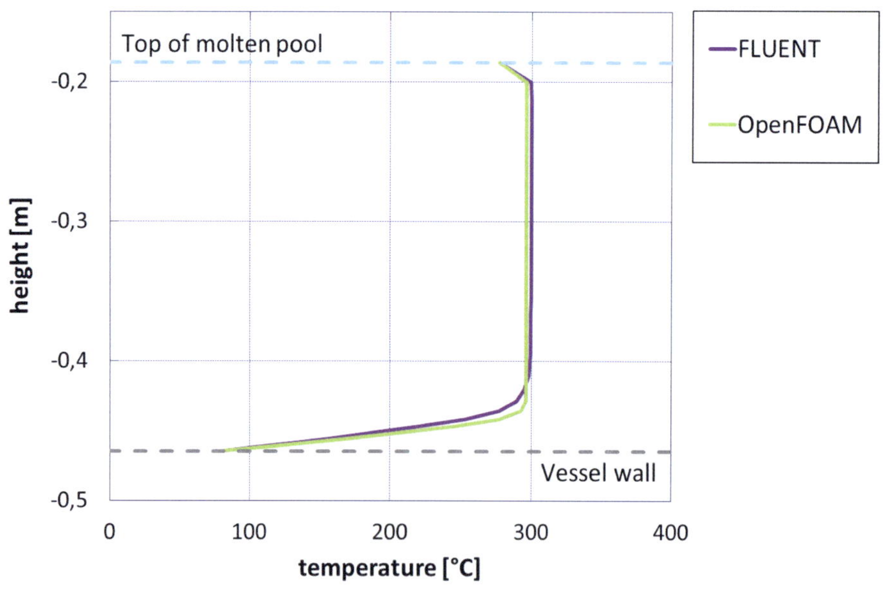

Figure 3.11: Temperature profiles in the melt pool along a vertical line at a distance 0.175 m from the symmetry axis

The temperature profile in figure 3.12 along a horizontal line at a distance of 0.17 m from the bottom of the lower plenum shows the same slightly lower mean temperature already observed in the vertical profile. The temperatures at the wall as well as the entire profile are again in very good concordance.

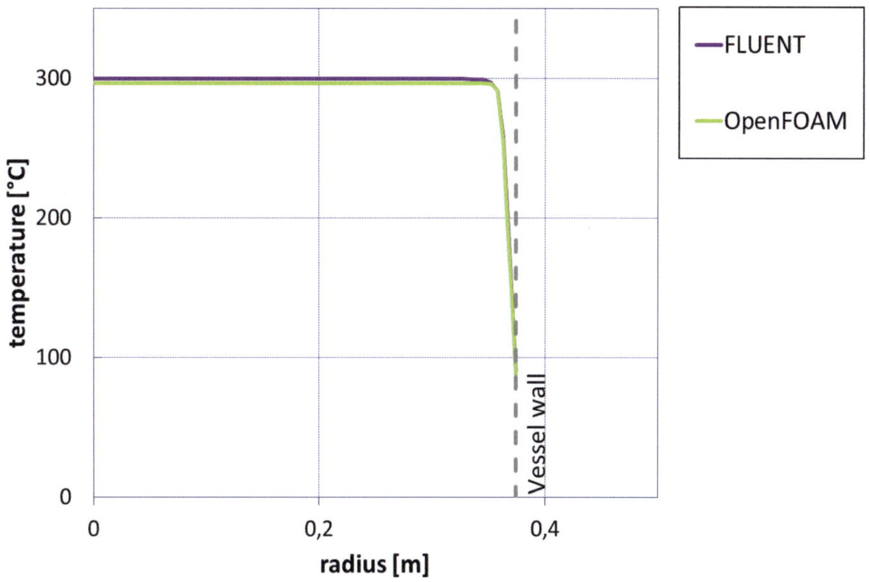

Figure 3.12: Temperature profiles in the melt pool along a horizontal line at a distance of 0.17 m from the bottom of the lower plenum

Figure 3.13 shows the temperature distribution in the entire calculation domain in ANSYS FLUENT and OpenFOAM. A slight different the temperature distribution a the bottom of the lower plenum can be found. This is caused by the slightly higher temperature gradient in the OpenFOAM simulation. Despite this different behaviour the two temperature distributions are also match very well.

Based on the simulations results obtained from the calculation with the *PECM* in ANSYS FLUENT and OpenFOAM, the *PECM* was implemented correctly. The slight differences which were found in the temperature profiles are caused by the two codes's different approaches to solve the linear equation system and can be neglected.

3.4.2 Simulation of LIVE-L1 with OpenFOAM

In contrast to the calculations performed to verify the correct implementation of the *PECM* in OpenFOAM in the *LIVE*-L1 test the facility is closed by the insulation lid. The air between the molten material and the insulation lid has no significant effect on the cooling. Similar to the previous calculations at the outside of the *LIVE*-vessel a constant temperature is applied. At the top of the molten material, however, no coolant is applied. The numerical grid and the evaluation lines for the simulation were shown in the previous chapter in figure 3.9 and figure 3.10.

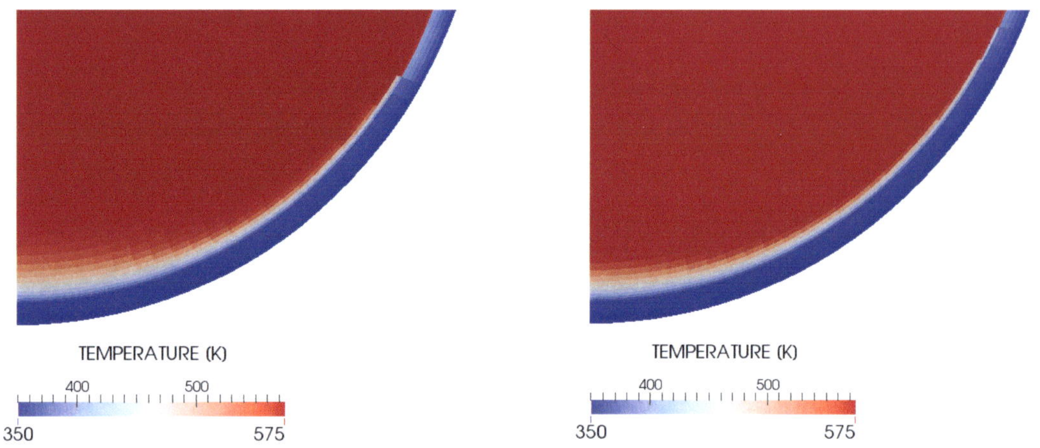

(a) Temperature in ANSYS FLUENT

(b) Temperature in OpenFOAM

Figure 3.13: Comparison of the steady state temperature distribution with the *PECM* in ANSYS FLUENT and OpenFOAM

Figure 3.14 shows the vertical temperature profile at a distance of 0.175 m from the centre of the facility in the *PECM* in comparison with the experimental data from *LIVE*-L1. The mean temperature is captured significantly better by the *PECM* in comparison to the stand-alone *MELCOR* calculation (ref. figure 3.7). However, towards the lower head at the bottom the stratification is not resolved.

The simulation with the *PECM* allows for the visualisation of the temperature distribution in the entire molten material. In figure 3.15 this temperature distribution for the steady state with 10 kW is shown.

The resolution of the temperature distribution allows for the monitoring of crust formation processes. The crust in the investigated steady state corresponds to the blue region in figure 3.16. The red region represents the entirely molten material in the *LIVE*-facility. The region between the crust and the fully molten material is the mushy zone, which is caused by the non-eutectic solidification process of the salt mixture (ref. figure 2.4).

The simulation of *LIVE*-L1 showed that the *PECM* in OpenFOAM is capable of providing a more detailed description of the molten material in the experimental facility compared to

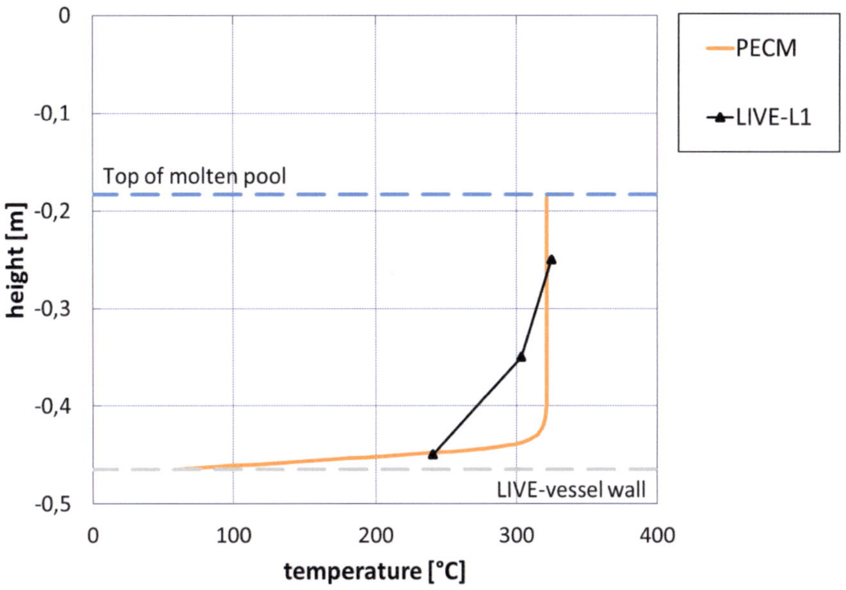

Figure 3.14: Temperature profiles in the melt pool along a vertical line at a distance 0.175 m from the symmetry axis compared to *LIVE*-L1 results

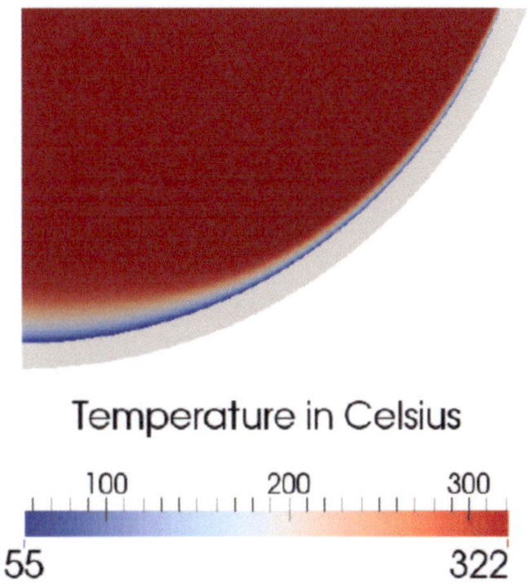

Figure 3.15: Calculated temperature distribution in the salt melt in *LIVE*-L1 with the *PECM*

the stand-alone simulation in *MELCOR*. The *PECM* provides a better agreement with the experimental findings and also allows for the visualisation of the temperature distribution within the entire molten material. In comparison to the stand-alone *MELCOR* simulation, which is unable to predict a crust along the lower head, the simulation in OpenFOAM can resolve the crust formation process as well as the non-eutectic behaviour of the molten salt mixture. These advantages qualify the *PECM* to be used in cases in which a more detailed simulation of the molten material in the lower head during a severe accident is necessary. In the following chapter the *PECM* in OpenFOAM will be used to simulate the experiment *LIVE*-L7v, in which another cooling concept for the salt mixture is applied.

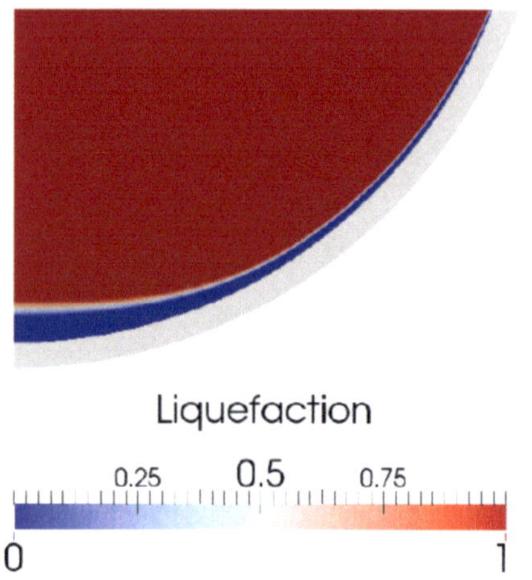

Figure 3.16: Liquid (red) and solid (blue) salt in *LIVE*-L1 calculated by the *PECM*

4 Expansion of the Phase-Change Effective Convectivity Model

This chapter will present the expansion of the Phase-Change Effective Convectivity Model (*PECM*) taking into account the stratified region in molten material which is heated internally and cooled through all adjacent surfaces. According to the description in chapter 2.2.4 the *PECM* treats the entire molten core material as a turbulent mixed zone. However, in an internally heated fluid, which was discussed in detail in chapter 2, different thermal regimes can be present. Figure 1.1 on page 4 showed these different flow regimes, which can occur in a pool of molten material in the lower plenum cooled from the top and through the side walls.

In the upper part of the fluid the assumption of a turbulent mixed region with a uniform temperature in the *PECM* is correct. Towards the lower boundary a stratification of the temperature profile can be found. The *PECM* cannot take this stratification into account, which was demonstrated by Chi-Thanh [15]. He compared the *PECM* to the results of a Computational Fluid Dynamics (*CFD*)-calculation. In contrast to the *PECM*, the *CFD*-calculation was able to resolve the stratified zone in the lower part of the molten material.

In this chapter first the simulations of the *LIVE*-L7v test with the traditional *PECM* approach will be discussed. The new approach, which takes into account the stratified region in the lower part of the molten material, will be developed based on the results obtained from these simulations. Afterwards, the Stratified Approach for the Phase-Change Effective Convectivity Model (*S-PECM*) will be verified by the recalculation of different states of the *LIVE*-L7v test.

4.1 Simulation of LIVE-L7v with the PECM

In the experiment series *LIVE*-L7 [52] the behaviour of internally heated material in the lower plenum, which is cooled from the top and the side walls, is investigated. To enable the cooling from the top the insulation lid used in the experiments L1, L10 and L11 is replaced by a cooling lid, which is placed directly on top of the molten material. During the experiment this lid can be cooled by water. In the *LIVE*-L7v test the salt mixture presented in chapter 2.2.2 is used to simulate the molten core material (Gaus-Liu et al. [61]). In this test different heating powers were applied. A steady state was observed at 9, 18, 24 and 29 kW heating power, respectively, generated by the volumetric heating system of the facility. The simulation setup is shown in figure 4.1.

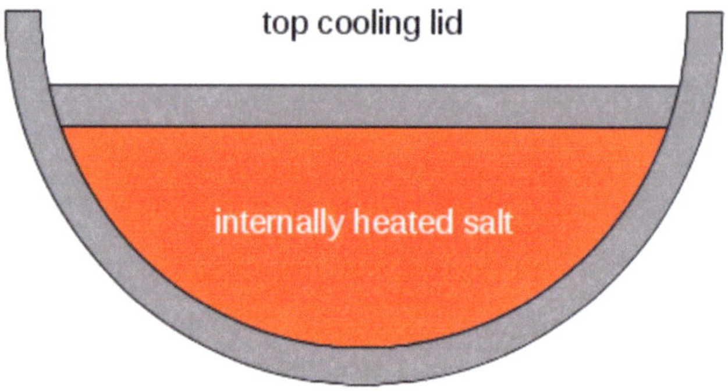

Figure 4.1: Simulation setup for the *LIVE*-L7v test with the *PECM*

As a calculation domain the numerical grid used for the simulation of the *LIVE*-L10 and *LIVE*-L11 tests (ref. figure 5.13) is used due to the same height of the molten material in the experiment. In *LIVE*-L7v a constant coolant support for the top-cooling lid as well as the external cooling of the vessel is established. Based on the experimental values a uniform temperature for the external coolant was applied. At the top boundary due to the experimental data a constant temperature of 102 °C is assumed. The temperature profile along the vertical line at a radius of 0.185 m from the centre of the *LIVE*-facility for the steady state in *LIVE*-L7 with a heating power of 29 kW simulated with the *PECM* is shown in figure 4.2.

The *PECM* significantly overestimates the temperatures monitored in the experiment. Also, the stratified regime at the bottom of the melt pool is not resolved. A similar result is achieved for the steady states with a volumetric heating power of 18 and 24 kW. The vertical temperature profiles for these steady states are shown in figure 4.3.

Despite the more accurate prediction of the temperature in *LIVE*-L7v for the volumetric heating powers of 18 and 24 kW, the stratified part towards the bottom of the lower plenum cannot be resolved by the *PECM*. For the steady state observed with a heating power of 9 kW in the *PECM* no molten material is calculated and a diffusive heat transfer in the entire debris in the lower plenum is assumed.

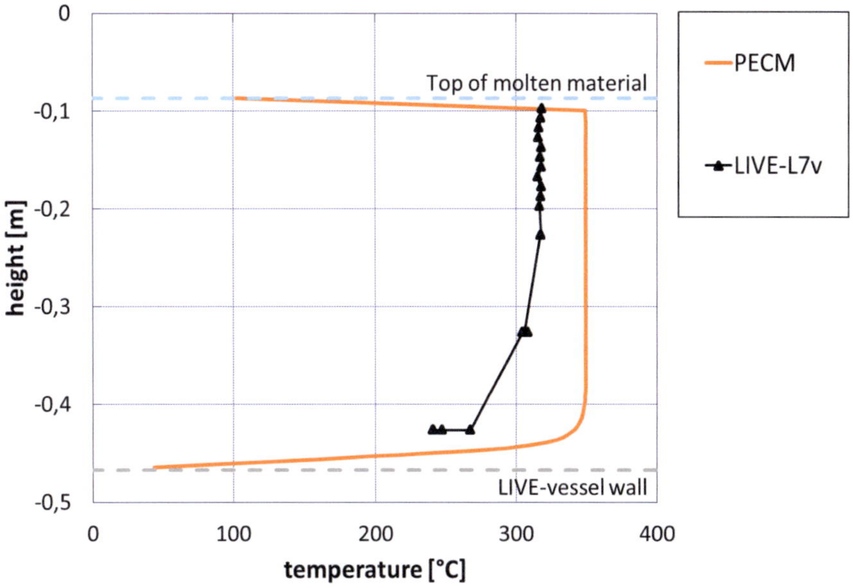

Figure 4.2: Vertical temperature profile at 0.185 m from the centre with a volumetric heating power of 29 kW in the *PECM* compared to experimental data from *LIVE*-L7v

4.2 Stratified Approach of the PECM

The *PECM* was modified in order to reduce the deviances in the temperature field of the *PECM*, compared to the experimental results. In the methodology of the *PECM* presented in chapter 2.2.4 the energy introduced to the system by the convective source term is subtracted uniformly from each cell to preserve the energy conservation. In the mushy-zone merely a part of the cell is molten (ref. figure 2.5) and therefore it represents only a part of the pool. In a first approach the artificial heat sink of the *PECM* is applied according to the percentage of liquid material in the cell. The energy added to the system by the convective source in the *PECM* is still reduced in total. Now, the distribution of the heat sink is no longer uniform. The temperature profile along the vertical line at 0.1875 m from the centre compared to the experimental findings of *LIVE*-L7v and the *PECM* with a volumetric heating power of 29 kW is shown in figure 4.4.

The reduction of the artificial heat sink in the mushy-zone according to the amount of solid material in the cell, reduces the deviance between the temperature found in the experiment and the one calculated by the *PECM*. The stratification in the bottom part of the molten material is still not resolved in the modified model. A comparison of the temperature distributions in the lower plenum between the traditional and the modified *PECM* is illustrated in figure 4.5.

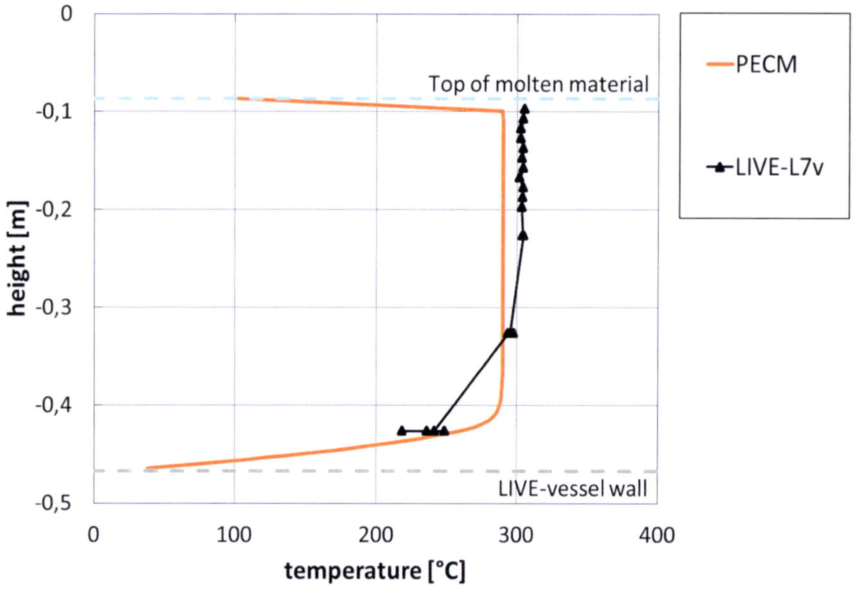

(a) Steady state with 18 kW volumetric heating power

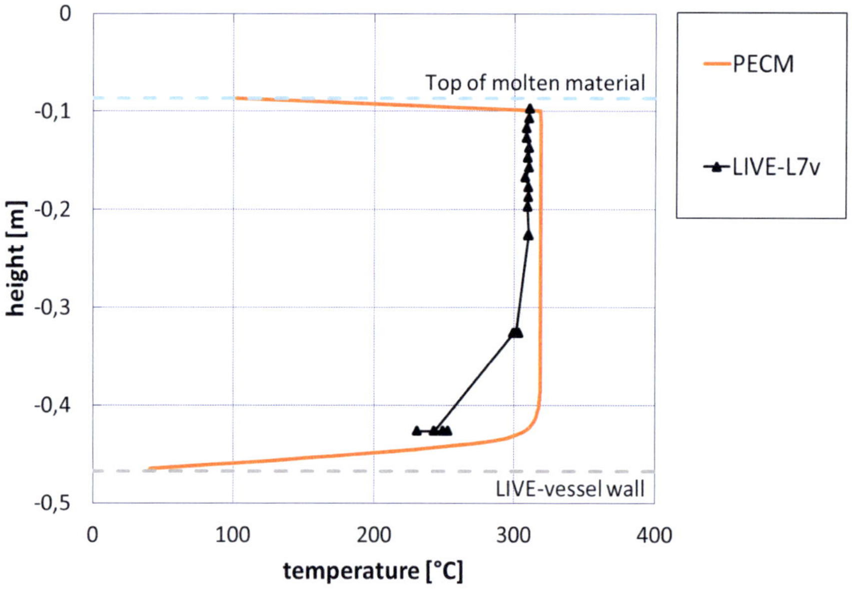

(b) Steady state with 24 kW volumetric heating power

Figure 4.3: Vertical temperature profiles at 0.1875 m from the centre in the *PECM* compared to experimental data from *LIVE*-L7v

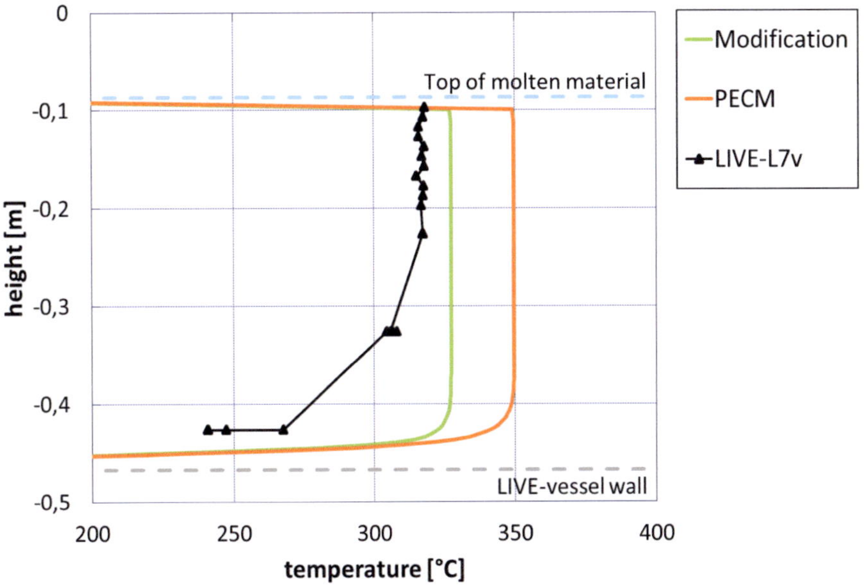

Figure 4.4: Vertical temperature profile at 0.185 m from the centre with a volumetric heating power of 29 kW in the modified *PECM* compared to experimental data from *LIVE*-L7v and the traditional *PECM*

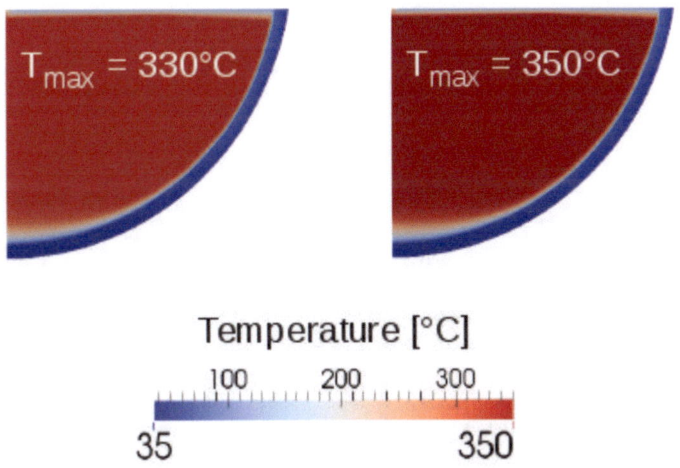

Figure 4.5: Temperature distribution in the modified *PECM* (left) compared to the traditional approach (right)

Willschütz et al. [21] use an approach similar to the *PECM*. The Nusselt number is calculated using the physical and geometric properties of the molten material. In contrast to the *PECM* the Nusselt number is not used to determine a characteristic velocity for the convective heat transfer. The thermal conductivity of the molten pool is transferred into an effective conductivity by a multiplication with the determined Nusselt number. In the approach of Willschütz et al. there are two different zones to be distinguished. The turbulent mixed zone, where the effective conductivity affects the heat transfer in all directions resulting in a homogeneous distribution of the temperature. In the lower part, where the pool of molten material shows a stratification, the effective conductivity affects the horizontal heat transfer.

The results of Willschütz et al. show that this approach allows for the resolution of the stratified region in the lower part of the molten pool. In order to use this approach prior to the simulation, however, the stratified and turbulent mixed regions have to be determined. These two regions cannot always be distinguished at the beginning of the simulation due to transient behaviour of the pool. In the *PECM* the division of the molten core material into an upper and lower part is dependent on the current pool geometry. This classification can be used to determine the regions of the upper and lower pool in each time step of the simulation allowing for a transient adaption of the heat transfer mechanism. For the lower part of the pool, as determined by the *PECM*, the horizontal thermal conductivity is modified by a multiplication with the Nusselt-number for the sidewards heat transfer.

$$\lambda_{\text{effective, horizontal}} = \lambda_{\text{horizontal}} \cdot Nu_{\text{sidewards}} \tag{4.1}$$

In the upper part of the molten material the traditional approach of the *PECM* presented in chapter 2.2.4 using the characteristic velocities to determine the temperature distribution and heat transfer. This is realised by a blending factor α_{blend}, which is 1 in the turbulent mixed regime and 0 in the stratified part of the pool.

$$\text{Model}_{\text{Heat Transfer}} = \alpha_{blend}(\text{PECM}) + (1 - \alpha_{blend})(\text{Effective Conductivity}) \tag{4.2}$$

A parameter study is performed to determine the correct position of the height dividing the upper and lower pool. Therefore, the height proposed by the *PECM* is multiplied with an external factor. The vertical temperature profiles for selected factors are shown in figure 4.6.

The effective conductivity approach by Willschütz et al. [21] introduces a stratified temperature profile in the bottom part of the molten material. The temperature profiles with a factor of 1.0 and 0.8 show a significant decrease of the temperature resulting in a concave temperature

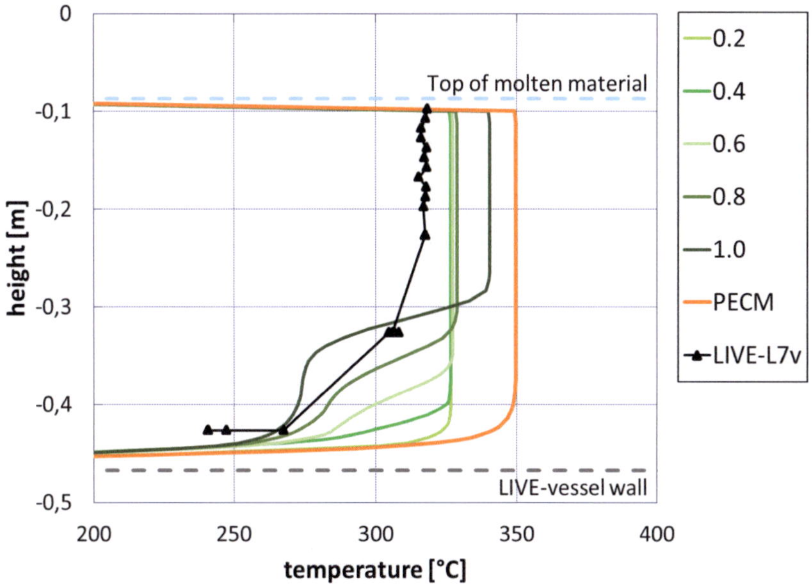

Figure 4.6: Vertical temperature profiles at 0.185 m from the centre with a volumetric heating power of 29 kW at different elevations for the stratified zone compared to experimental data from *LIVE*-L7v and the traditional *PECM*

profile. Furthermore, an increased temperature in the upper part compared to the smaller factors can be observed. With the factor 0.2 no significant temperature difference can be monitored compared to the modified *PECM* in figure 4.4. The temperature distribution in the lower plenum for these different factors is shown in figure 4.7.

With a factor of 1.0 and 0.8 the sharp reduction of the temperature in the stratified part can be observed in the temperature distribution of the lower part. A factor of 0.2 leads to a slightly higher temperature compared to the temperature without a stratification region. The lowest temperature combination with a significant stratified region can be found when a factor between 0.4 and 0.6 is applied. Further investigations in this region showed that the optimal factor is independent of the applied heating power and depends on the geometric shape of the calculation domain. The determined factor matches the geometric relation between the area of the circle at the determined height of the lower pool in the *PECM* A_{circle} and the surface of the corresponding circular segment at this height A_{segment} subtracted from 1.

$$factor = 1 - \frac{A_{\text{circle}}}{A_{\text{segment}}} \tag{4.3}$$

For a hemispherical lower head this factor always shows a value between 0.67 and 0.5. This factor physically represents the relationship between the heat transferred to the top of the lower region, which is governed by the *PECM* and the heat removed through the adjacent walls of the lower part. A scheme of the different heights and surfaces used to determine this factor can be found in figure 4.8.

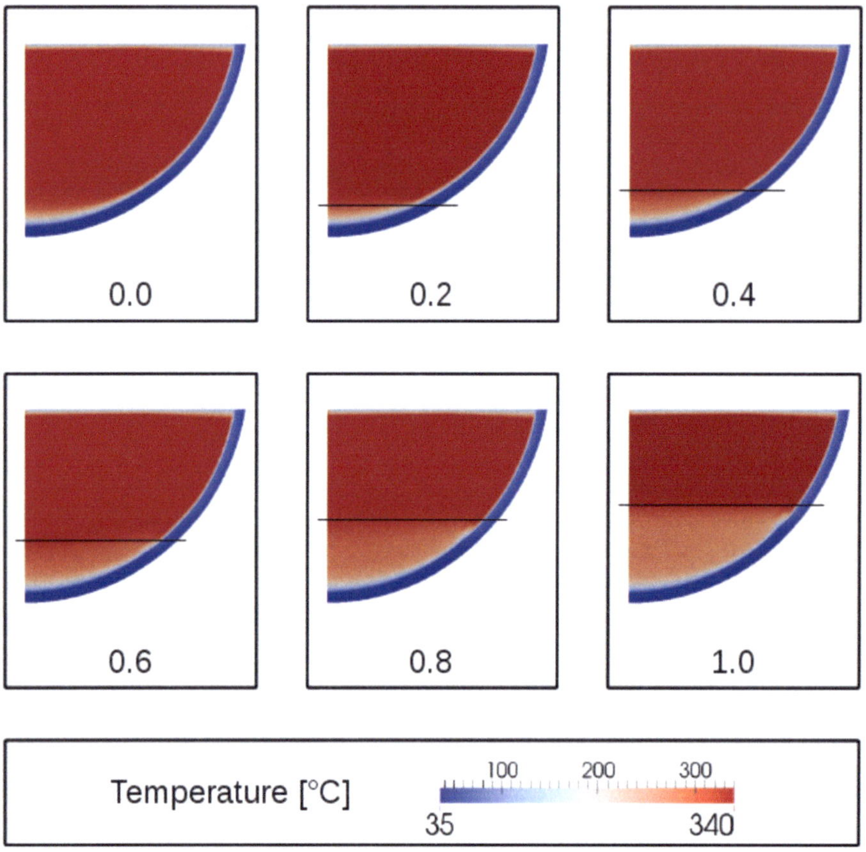

Figure 4.7: Temperature distribution with a volumetric heating power of 29 kW at different elevations (black lines) for the stratified zone

4.3 Simulation of LIVE-L7v with the S-PECM

The determined splitting factor is implemented into the source code of the *PECM*. The modified *PECM*-approach allows for the calculation of stratified parts in the molten material. This new developed Stratified Approach for the Phase-Change Effective Convectivity Model (*S-PECM*) is used to recalculate the different steady states found in the *LIVE*-L7v experiment. The vertical temperature profile for the steady state with 29 kW at a radius of 0.1875 m from the centre of the facility is shown in figure 4.9 compared to the experimental findings and the traditional *PECM*-approach.

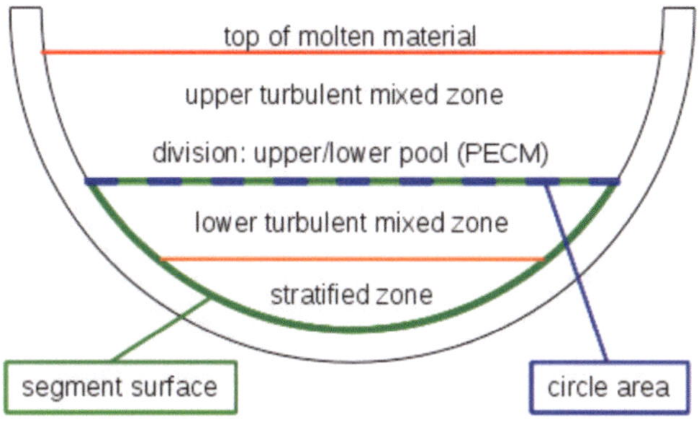

Figure 4.8: Surfaces to determine the height of the stratified region in a hemispherical lower head

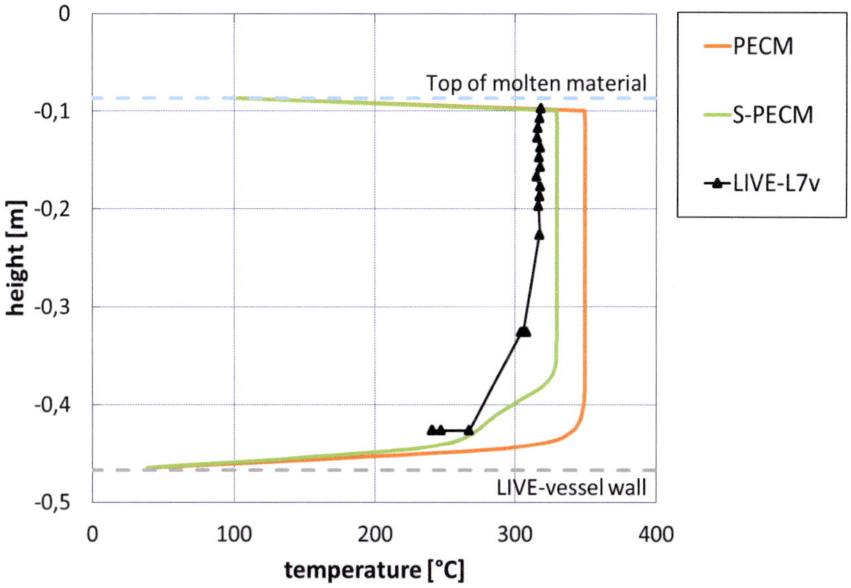

Figure 4.9: Vertical temperature profile in the *S-PECM* at a distance of 0.1875 m from the centre with a volumetric heating power of 29 kW compared to experimental data from *LIVE*-L7v and the traditional *PECM*

In contrast to the traditional *PECM*-approach the *S-PECM* simulation shows a significant reduction of the mean temperature, which is in better allignment with the experimental findings. Additionally, the *S-PECM* calculates a stratified region in the lower part of the molten material. Compared to the previous simulation results calculated by the traditional *PECM* the modified approach allows for a more precise and realistic simulation of the molten material in the lower plenum. A comparison of the temperature distribution in the lower plenum between the *PECM* and the *S-PECM* approach can be found in figure 4.10.

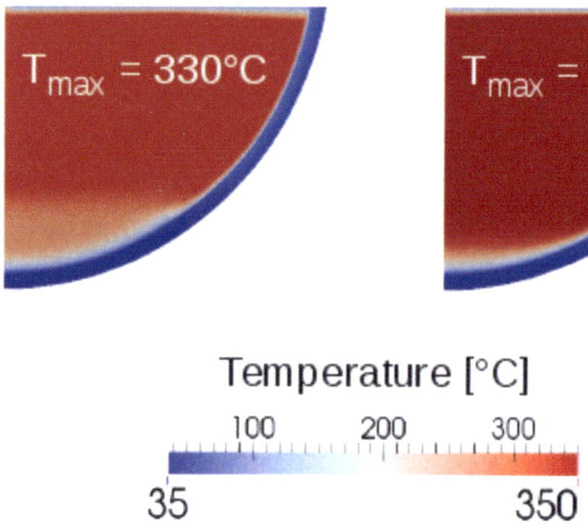

Figure 4.10: Temperature distribution in the *S-PECM* (left) compared to the traditional *PECM*-approach (right)

The reduced temperature in the turbulent mixed region in the upper part of the pool in the *S-PECM* compared to the traditional *PECM* can also be clearly observed in the illustration of the temperature distribution. Furthermore, in the lower part the stratified temperature profile can be identified. The different temperature profiles and distribution in the *S-PECM* also effects the crust along the inner lower head surface. For the volumetric heating power of 29 kW the crust thickness in dependence of the lower head angle in the *S-PECM* is shown in figure 4.11.

In the stratified region the crust formed at the lower head is significantly thicker in the *S-PECM* than the crust calculated by the traditional *PECM*-approach. The trend of the experimental data for the crust thickness in the upper part of the vessel shows an increase of the crust thickness towards the lower parts of the vessel. Also the measurement point at about 37° shows a thicker crust than predicted by the traditional *PECM*. Therefore, the crust formed in the *S-PECM* can be assumed to be more precise. The effects of the *S-PECM* on the calculation results can be observed in the steady states of the *LIVE*-L7v test with heating powers of 18 and 24 kW. The vertical temperature profile at 0.1875 m from the centre of the *LIVE*-facility in the *S-PECM* for

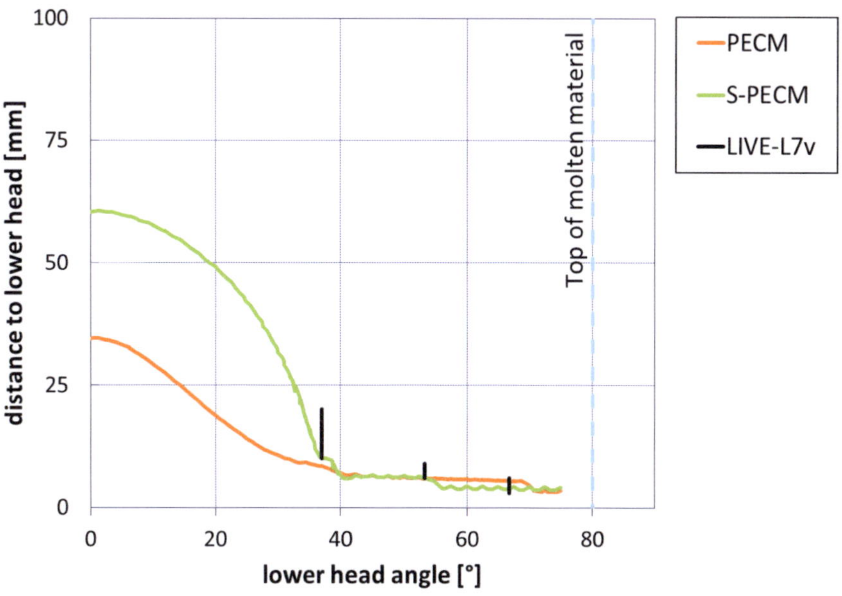

Figure 4.11: Crust thickness along the lower head in the *S-PECM* with a volumetric heating power of 29 kW compared to experimental data from *LIVE*-L7v and the traditional *PECM*

these heating powers compared to the traditional *PECM*-approach and the experimental data is shown in figure 4.12.

In both steady states a more precise and realistic prediction of the temperature profile was calculated by the *S-PECM* compared to the traditional *PECM*. In the steady state with a heating power of 24 kW the turbulent mixed temperature is reproduced exactly by the *S-PECM*. Both simulations with the *S-PECM* also show a stratification of the melt pool in the lower part, which is determined by the factor introduced in equation 4.3. Compared to the experimental data there are still deviances in the stratification height. Compared to the traditional *PECM*, however, the stratification and the reduction of the temperature in the turbulent mixed region is a significant progress. The different temperature profiles and distribution in the entire lower plenum, also effect the crust formed along the lower head. For the two steady states of *LIVE*-L7v with volumetric heating powers of 18 and 24 kW the formed crust is compared to the experimental data and the original *PECM* in figure 4.13.

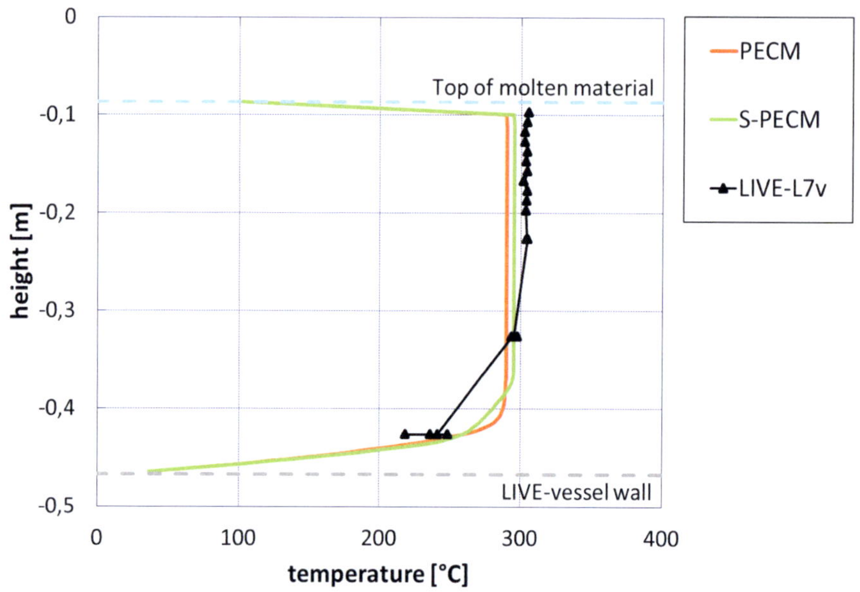

(a) Steady state with 18 kW volumetric heating power

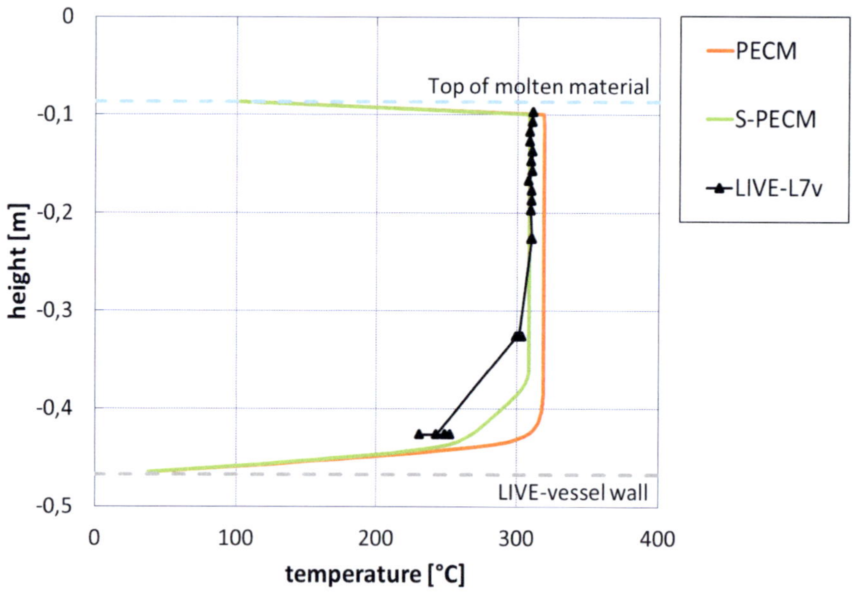

(b) Steady state with 24 kW volumetric heating power

Figure 4.12: Vertical temperature profiles at 0.1875 m from the centre in the *S-PECM* compared to experimental data from *LIVE*-L7v and the traditional *PECM*-approach

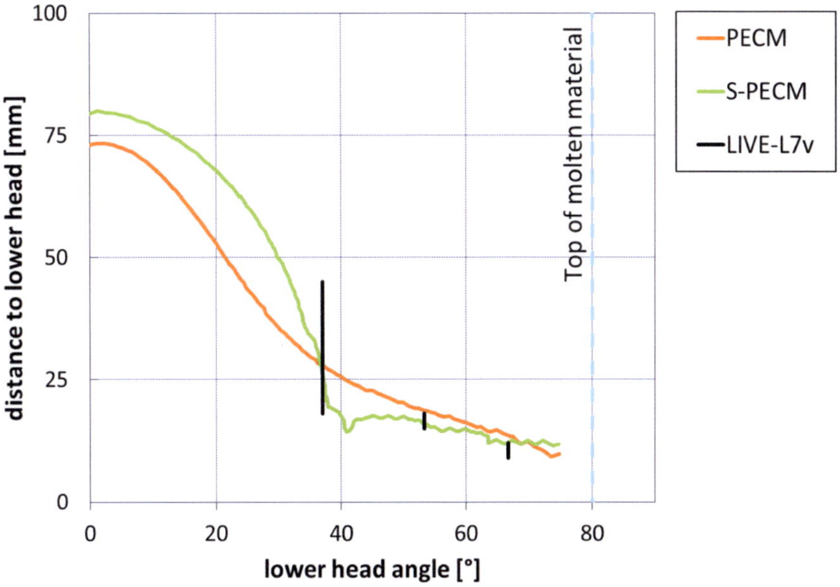

(a) Steady state with 18 kW volumetric heating power

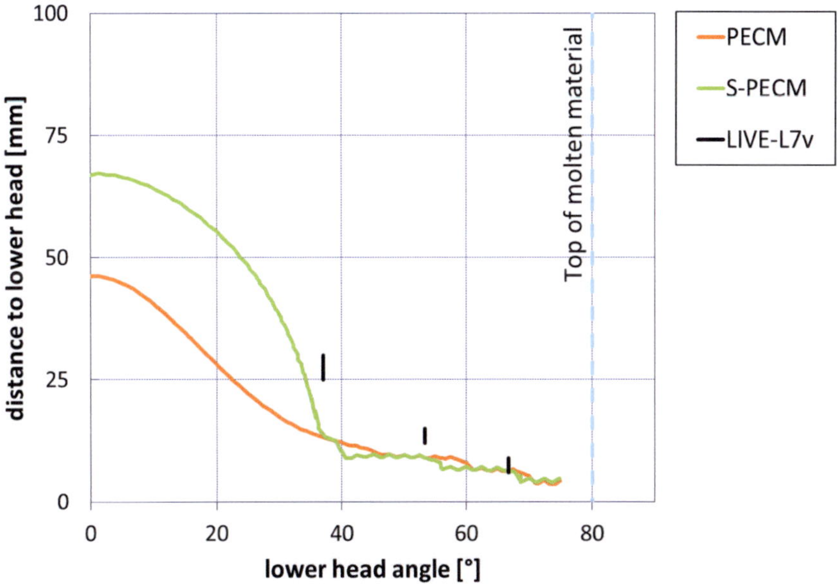

(b) Steady state with 24 kW volumetric heating power

Figure 4.13: Crust thickness along the lower head in the *S-PECM* compared to experimental data from *LIVE*-L7v and the traditional *PECM*-approach

For a heating power of 18 kW the crust thickness in the *S-PECM* is similar to the traditional *PECM*. The experimental values show that in the upper part of the vessel the crust is slightly overestimated by the *PECM*. The *S-PECM* matches very well with the experimental values. In simulations with a heating power of 24 kW both the *S-PECM* and the traditional *PECM* underestimate the crust thickness. Therefore, the S-*PECM* shows a significantly thicker crust in the lower part of the vessel due to the stratified region which is in better accordance with the trend from the *LIVE*-L7v test.

Especially for a volumetric heating power of 29 kW,the simulations of *LIVE*-L7v with the *PECM* showed the overestimation of the temperature in the turbulent mixed region. A modification of the artificial energy sink in the *PECM* significantly improves the prediction of the temperature in the turbulent mixed region. However, both the traditional and the modified *PECM* cannot reproduce the stratification of the molten core material in the lower part of the pool. The *PECM* divides the pool into an upper and a lower part but assumes a turbulent mixed regime in both of them. The approach of Willschütz et al. [21], who modified the thermal conductivity of the molten material according to the predicted Nusselt-number, was applied for the lower pool region in the *PECM*. A parameter study was performed to determine the part of the lower pool region which is treated by the modified conductivity. According to this study the best results were obtained when about 50 % - 70 % of the lower pool regions height is treated by the approach of Willschütz et al.. The rest of the molten material is assigned to the turbulent mixed region, which is treated by the traditional *PECM*-approach. The factor extracted from the parameter study depends on the geometric properties of the surrounding geometry and can be dynamically determined during the calculation using equation 4.3. With the stratified approach three different steady states in the *LIVE*-L7 experiment were simulated. In comparison to the traditional *PECM* in the *S-PECM* the temperatures of the turbulent mixed zone are in significantly better accordance with the temperatures measured in *LIVE*-L7v. Furthermore, with the *S-PECM* the prediction of a stratified region in the lower part of the molten material is possible without the necessity of dividing the calculation domain into different regions prior to the execution of the simulation. The prediction of the crust thickness along the lower head was also improved compared to the traditional *PECM*, which underestimates the crust thickness in most cases. Even with the *S-PECM* there are still deviances in the temperature profile as well as in the crust thickness compared to the experimental data. However, in comparison to the traditional *PECM* with the *S-PECM* a more precise and realistic simulation of internally heated molten material which is cooled by all of the surrounding structures, is possible in a similar amount of calculation time.

5 Coupled simulations of the LIVE-facility

Chapter 3 showed that the Phase-Change Effective Convectivity Model (*PECM*) has the ability to perform a significantly more detailed simulation of the behaviour of molten core material in the lower plenum of a Reactor Pressure Vessel (*RPV*) than a standalone *MELCOR* simulation In order to improve the prediction of an entire severe accident in a Nuclear Power Plant (*NPP*) with corium in the lower plenum, the *PECM* is coupled to the system code *MELCOR*. This coupling allows for the use of the more precise *PECM* in *MELCOR* and also provides boundary conditions from the *NPP*-simulation in *MELCOR* for the calculations in OpenFOAM with the *PECM*.

In this chapter the coupling methodology developed in this work will be presented. This coupling between *MELCOR* and the *PECM* in OpenFOAM shall be used to simulate the experiment *LIVE*-L1, which has already been simulated by the stand-alone versions of *MELCOR* and the *PECM* in chapter 3. In order to provide additional validation data for the *PECM*, *MELCOR* and the developed coupled system additionally the experiments *LIVE*-L10 and *LIVE*-L11 will be simulated. In these experiments the molten salt mixture in the facility is not cooled by the cooling lid from the top. Therefore, in this chapter the original approach of the *PECM* will be used in the coupled simulations.

5.1 The coupling tool DINAMO

In order to couple new models to the system code *MELCOR* an external tool called *DINAMO* (Direct Interface for Adding Models) was developed. This program consists on the one hand of the coupling routines to enable the communication via the coupling interface with *MELCOR* via the MPIEXEC program (ref. chapter 2.4.2). On the other hand *DINAMO* provides an interface which allows for the integration of new models into the source code as well as an easy communication with external programs. The basic structure and routines of *DINAMO*, which are entirely written in FORTRAN-90, are shown in figure 5.1.

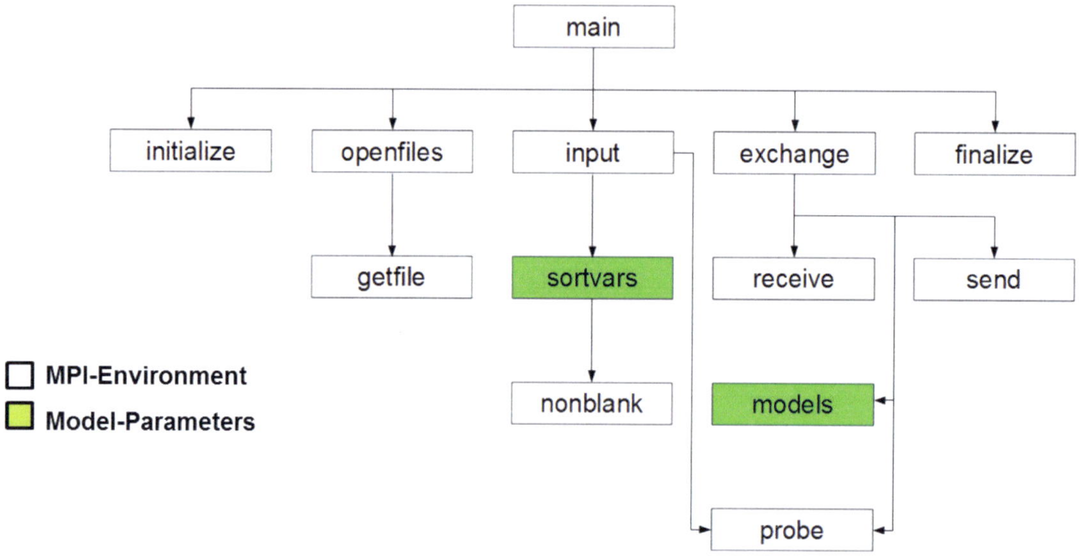

Figure 5.1: Basic structure of the program *DINAMO*

The routines highlighted in green are used to create the interface for the communication with implemented models or external programs. The white routines are based on the coupling interface in *MELCOR* and enable the communication to the coupled system code via MPIEXEC. Appendix A presents the general program execution during a coupled simulation as well as a detailed discussion of the model interface in *DINAMO*.

5.2 Test of the program DINAMO

This chapter covers the calculations performed to test the correct functionality of the developed coupling tool *DINAMO*. The Larson-Miller-Model, which allows for the prediction of structure failure due to creeping, is implemented in *DINAMO* and used to calculate the failure of the lower core support plate. This chapter will first introduce a brief overview of the Larson-Miller-Model. Afterwards the implementation of this model into *DINAMO* and the coupled simulation results will be presented.

5.2.1 The Larson-Miller-Model

In *MELCOR* the failure of the lower head can be calculated by the Larson-Miller-Model [93]. This model takes into account the stress caused by the relocated material to the lower plenum and the temperature of the structure [73]. In order to determine if the structure fails due to temperature and stress a Larson-Miller-Parameter P_{LM} is calculated.

$$P_{LM} = C_1 log(\sigma_{eff}) + C_2 \tag{5.1}$$

σ_{eff} is the stress on the structure caused by the weight of the molten material in the lower plenum. C_1 and C_2 are constants, which are defined by the material of the structure. Using the Larson-Miller-Parameter, the temperature of the structure T and an additional material constant C_3 the remaining time until the failure of the structure t_R can be determined.

$$t_R = 10^{\left(\frac{P_{LM}}{T} - C_3\right)} \tag{5.2}$$

The remaining time until failure t_R specifies the length of time the intact structure could withstand the constant stress σ_{eff} and the constant temperature T. This assumption is based on experimental investigations and is called the stationary creep rate. The steady creep rate is not valid at the start of the stress on the structure and close to its failure. However, for the purposes of *MELCOR* it is approximated that the steady creep rate is valid from the start of the stress until the failure of the structure. This approximation allows for the calculation of the so called life time progress ε_{new} based on t_R according to the following equation:

$$\varepsilon_{new} = \varepsilon_{old} + \frac{\Delta t}{t_R} \tag{5.3}$$

The fraction of the current time step Δt from the time of t_R can be considered to be the fraction of damage caused in the structure by the current stress and temperature. ε_{old} is the life time progress calculated in the previous time step. If ε_{new} exceeds the value of 1, the structure fails.

5.2.2 Coupling the Larson-Miller-Model

The Larson-Miller-Model, which is already implemented in *MELCOR* to calculate the failure of the *RPV* lower head due to creeping, is implemented in *DINAMO* to test the coupling of external models via the modified coupling methodology. The coupled Larson-Miller-Model is used to calculate the failure of the lower core support plate. In order to externally use the Larson-Miller-Model the temperature and the current stress in the structure caused, for example, by relocating debris or the structure's own weight, has to be sent from *MELCOR* to *DINAMO*. In *DINAMO* these parameters are then used to calculate the life time progress ε_{new} as described in chapter 5.2.1. The result is sent back to *MELCOR*. A logical control function (ref. [93]) in *MELCOR* determines if the received ε_{new} exceeds 1. In this case the failure of the lower core support plate is initiated. Figure 5.2 shows a schematic illustration of the coupling process between *MELCOR* and the Larson-Miller-Model in *DINAMO*.

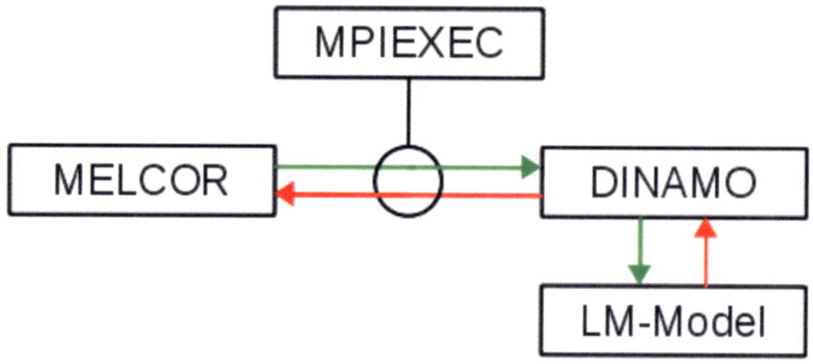

Figure 5.2: Schematic of the coupling process between to use the Larson-Miller-Model in *MELCOR*

In figure 5.3 the temperature profile in the lower core support plate in *MELCOR* and the coupled Larson-Miller-Model is compared to the standard *MELCOR* model with a failure at a temperature of approximately 1450 K.

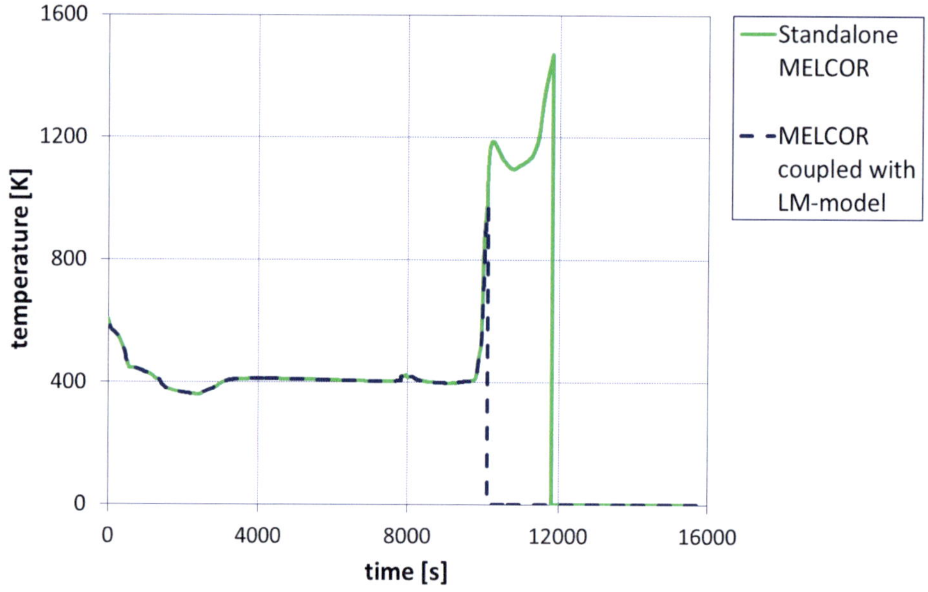

Figure 5.3: Temperature profile in the lower core support plate in *MELCOR* with and without the coupled Larson-Miller-Model

The failure of the lower core support plate in the Larson-Miller-Model occurs earlier than in the standard *MELCOR* model. The temperature profiles until the failure are exactly the same, which is the result of the synchronous coupling method (ref. appendix B). This calculation's purpose is to demonstrate the correct functionality of the synchronous coupling methodology.

Therefore, the deviances in the failure time are not discussed here. In the coupled simulation the failure is induced by the externally calculated life time progress ε_{new}. This coupled parameter and the value of the corresponding failure function in *MELCOR* are shown in figure 5.4

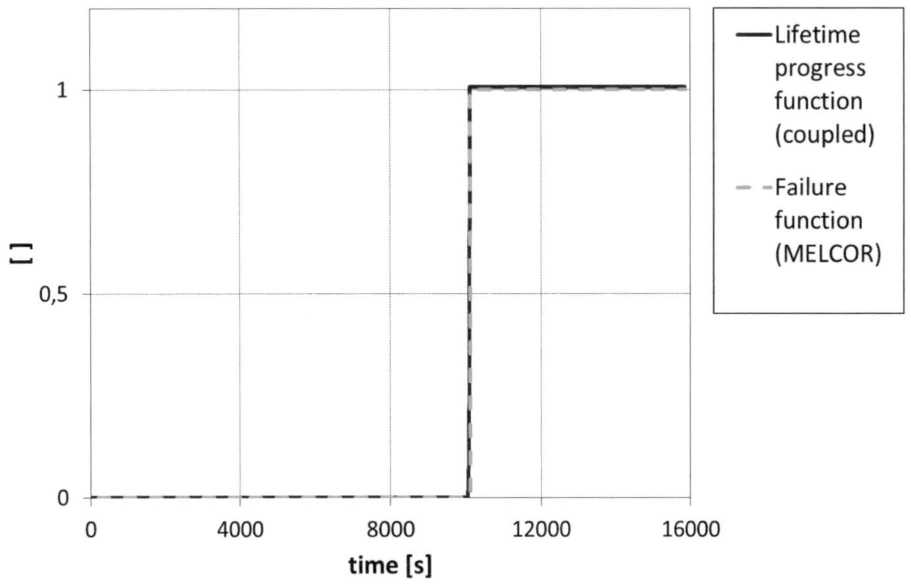

Figure 5.4: Progress of the life time progress function in *DINAMO* and the corresponding failure function in *MELCOR*

The sharp temperature gradient in the lower core support plate (ref. figure 5.3) causes the Larson-Miller-Model to predict a lifetime progress function in *DINAMO* higher than 1, which leads to the failure of the structure induced by the failure function in *MELCOR*.

The example of coupling the Larson-Miller-Model to *MELCOR* using *DINAMO* for the calculation of the failure of the lower core support plate shows that the modified coupling routine works properly. Also no influence of the coupling process on the simulation results could be found. Therefore, it is now possible to couple new and more complex models to *MELCOR* via the external tool *DINAMO*, thereby expanding *MELCOR*âĂŹs possibilities to predict a severe accident in an *NPP* without the necessity of a complex modification of the *MELCOR* source code.

5.3 Coupling the PECM with MELCOR

This chapter contains the work performed to couple the *PECM* in OpenFOAM (ref. chapter pecm-openfoam) with *MELCOR* to improve the prediction of molten core material in the lower plenum of the *RPV* of a *NPP* during a severe accident. Therefore, the interface for the communication with external models in *DINAMO* is modified to control the execution of OpenFOAM and manipulate the boundary conditions for the *PECM*-simulation based on the data received from *MELCOR*.

In order to simulate the behaviour of molten material in *MELCOR* using the *PECM* in OpenFOAM the coupling methodology shown in figure 5.2 has to be extended. The schematic illustration of this new coupling process is shown in figure 5.5. The interface for the implementation of new models in *DINAMO* was modified to enable the communication with OpenFOAM and to adress the solver with the *PECM*-routines in this Computational Fluid Dynamics (*CFD*)-code.

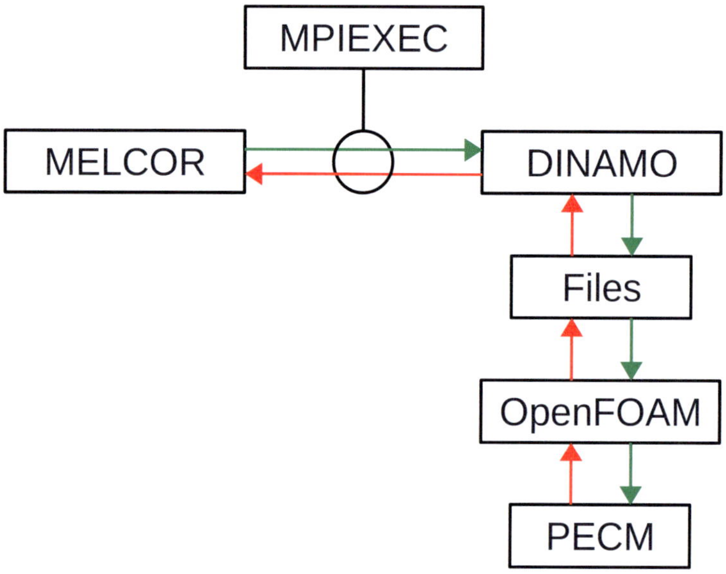

Figure 5.5: Scheme of the extended methodology to couple the *PECM* to *MELCOR*

The exchange of coupling parameters between *MELCOR* and *DINAMO* is executed in the same way as presented in chapter 2.4.2. Unlike the direct implementation of new models into the *DINAMO* source code (ref. chapter 5.2.2) the communication between *DINAMO* is based on external files. In one file the boundary conditions received from *MELCOR* are stored. This file used by OpenFOAM to establish the boundary conditions for the simulation of the next time step with the *PECM*-solver. At the end of each OpenFOAM time step the necessary data for the coupling exchange with *MELCOR* is stored into a separate external file. The data from this file is used by *DINAMO* to assign the corresponding boundaries in *MELCOR*. Afterwards these parameters are exchanged via the coupling interface using MPIEXEC. The execution and the time progress of the *PECM*-solver in OpenFOAM is controlled by *DINAMO* to enable a parallel simulation.

In order to perform a coupled simulation of the behaviour of molten core material in the lower plenum during a severe accident in an *NPP* or the behaviour of the molten salt in the *LIVE*-facility with the coupled system a numerical grid for the *PECM* has to be generated, in addition to the *MELCOR* input deck. This grid is used in OpenFOAM to solve the linear equation system of the *PECM*-solver.

At the beginning of a *MELCOR* simulation of a severe accident usually no molten core material is considered in the lower plenum. In this case, as already discussed in chapter 3.1, the simulation of the *LIVE*-facility does not start with molten material in the lower plenum. An activation function is introduced to the coupling parameters to indicate if sufficient molten material is available in the lower plenum to start the calculations in the *PECM*. After the activation of the *PECM* in OpenFOAM the simulation has to be initialised. The input data files of OpenFOAM are set to match the current environment simulated by *MELCOR*. The molten material inside the lower plenum is initialised using the mean melt temperature (*T_Melt*) from *MELCOR*. The performed coupling procedure is illustrated in figure 5.6.

During the run time of the simulation the boundary conditions such as the *MELCOR* temperature of the water coolant at the outside of the vessel (*T_Coolant*) and the temperature at the top of the molten material (*T_Top*) are written into special boundary condition files by *DINAMO*. This file is used by OpenFOAM to specify the boundary conditions for the current coupling time step. At the end of each coupling time step the temperatures along the inner lower head surface in the *PECM* are stored into an external data file. *DINAMO* uses these temperatures to determine the temperatures of the inner lower head surface in the *MELCOR* nodalisation by linear interpolation. These temperatures are sent to *MELCOR* using the coupling interface. In *MELCOR* this temperature profile is used to recalculate and adjust the temperatures of the inner lower head surface.

5.4 Expansion of the Validation-Data for the PECM

This chapter contains the coupled calculations performed on the *LIVE* facility to provide additional validation data for the *PECM*, *MELCOR* and the coupled system. The experiment *LIVE*-L1, which was already investigated by the stand-alone Methods for Estimation of Leakages and Consequences of Releases (*MELCOR*) and *PECM* simulations, is first recalculated by the coupled system. Subsequently, the simulations of the experiments *LIVE*-L10 and *LIVE*-L11 are discussed. In these experiments the influence of the external cooling conditions is investigated.

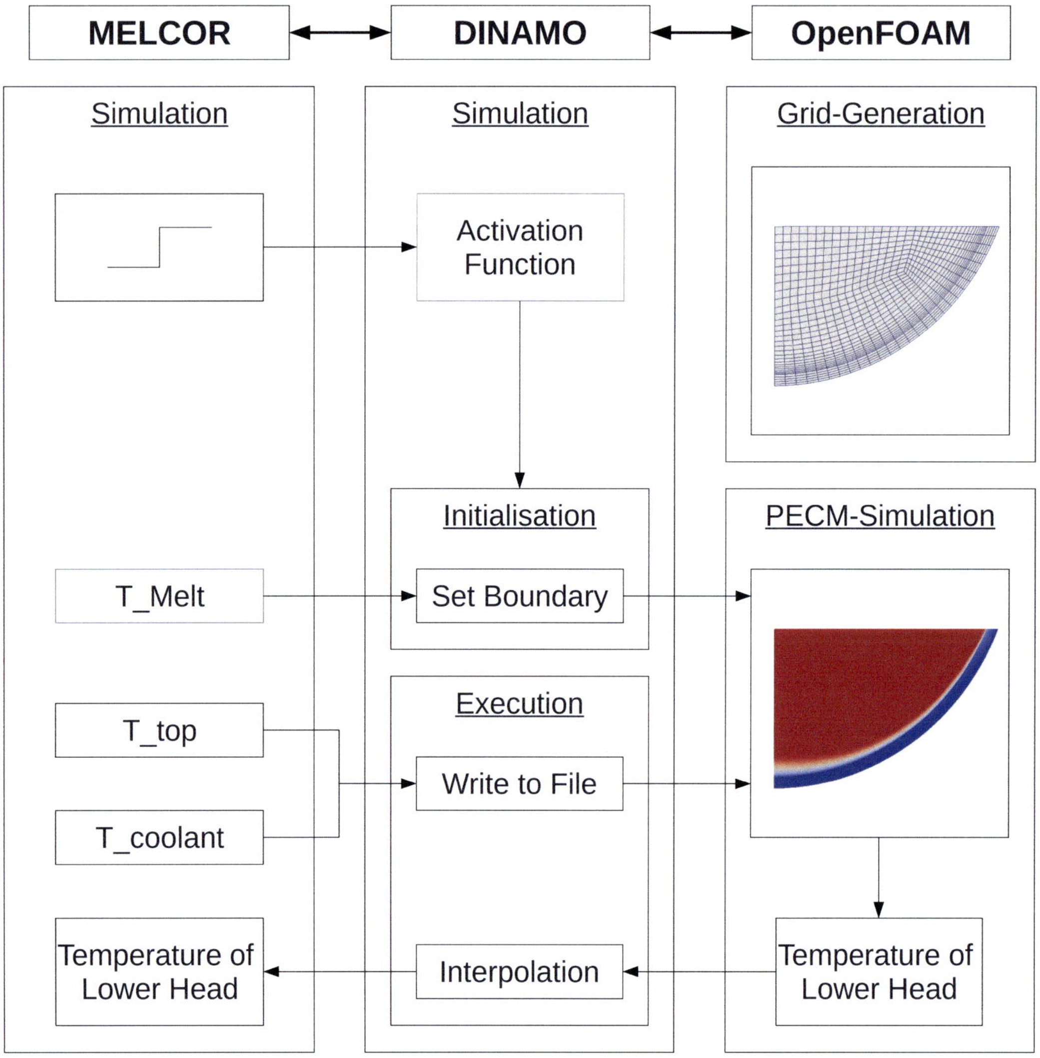

Figure 5.6: Coupling scheme between *MELCOR*, *DINAMO* and OpenFOAM for the implementation of the *PECM* to model molten core material in the lower plenum in *MELCOR*

5.4.1 Simulation of LIVE-L1

The first simulation with the coupled *MELCOR-PECM* system is carried out on the experiment *LIVE*-L1 [54]. In this experiment 120 l of molten salt is poured into the facility. This corresponds to a height of the molten material of approximately 31 cm. The volumetric heating system is used to heat up the salt at different heating powers. The *LIVE*-vessel is cooled by water from outside with a coolant support rate of about 10 cm^3/s at 23°C. The *LIVE*-facility is closed by the insulation lid. As observed by Fluhrer et al. [58] a stable steady state was observed at heating powers of 10 and 7 kW. In this chapter the simulation results of the two steady states are presented and compared with the experimental findings.

In the coupled simulation the calculations in the *PECM* are initialized by the activation function after all of the salt in *MELCOR* is liquefied and the flooding of the external coolant region begins (ref. chapter 3.3). Figure 5.7 shows the temperature distribution in the steady state calculated in OpenFOAM by the *PECM* in a coupled simulation. In OpenFOAM also a large region with a uniform temperature distribution can be found, which is the result of the assumption of a well mixed turbulent pool for the entire molten material. However, a temperature gradient is observed, especially close to the lower boundary of the *LIVE*-vessel.

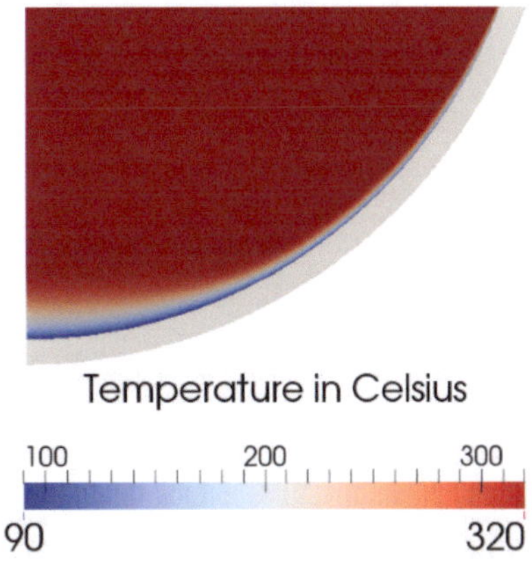

Figure 5.7: Temperature distribution for *LIVE*-L1 calculated by the *PECM*

The temperature of the inner vessel wall of the *LIVE*-facility is extracted from the *PECM*-simulation and sent to *MELCOR* via the coupling interface. Based on the coupled temperature profile the inner vessel wall in *MELCOR* is recalculated. In figure 5.8 the coupling domain with the active boundaries in the coupling process is highlighted.

The coupling of *MELCOR* with the *PECM* allows for the comparison of the local temperature in the *LIVE*-test with the simulation results obtained in OpenFOAM. This comparison between the vertical temperature profile in the *PECM* and the experimentally measured values in *LIVE*-L1 for the steady state with a volumetric heating power of 10 kW along a vertical line at 0.175 m from the symmetry axis of the calculation domain (ref. figure 3.10) is shown in figure 5.9. The assumption of a fully mixed molten salt pool in the *PECM* results in significant differences in the temperature profile towards the lower head. The deficiencies of the *PECM* to resolve this stratified region in the lower part of the vessel were monitored in the stand-alone *PECM*-simulation (ref. figure 3.14). This behaviour of the *PECM* was also documented by Tran [15], who compared the *PECM* to a traditional *CFD*-calculation. Despite these deviances the maximum temperature is captured by the *PECM*-calculations very well.

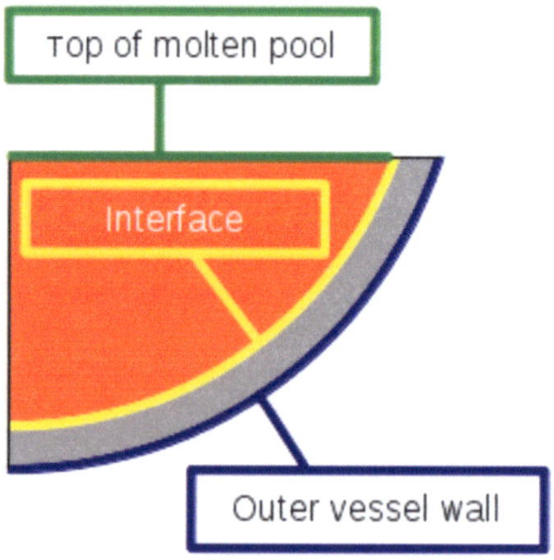

Figure 5.8: Active boundaries in the coupling process between *MELCOR* and the *PECM*

The prediction of the local temperature distribution in the molten material allows for the observation of the formation of a crust along the vessel wall of the *LIVE*-facility. The crust and mushy-zone thickness at different angles of the vessel wall are shown in figure 5.10.

The *LIVE*-L1 test ended after the steady state with a volumetric heating power of 7 kW was reached [58]. The liquid melt was poured from the facility back into the external furnace. The crust and mushy-zone, which were attached to the vessel wall, remained inside the facility and a detailed thickness measurement was performed. The comparison between the thickness of the remaining material in the *LIVE*-facility and the mushy-zone thickness calculated in the coupled *MELCOR*-*PECM* simulation is shown in figure 5.11. The remaining crust in the *LIVE*-facility is thicker than the calculated crust and mushy zone. The different temperature gradient close to the vessel wall in the *PECM* in comparison to the experiment can cause a higher temperature and a thinner crust at the lower head in the simulation. However, using the *PECM* a prediction of the crust formation is possible, which was not the case in the stand-alone *MELCOR* simulation of the *LIVE*-L1 test.

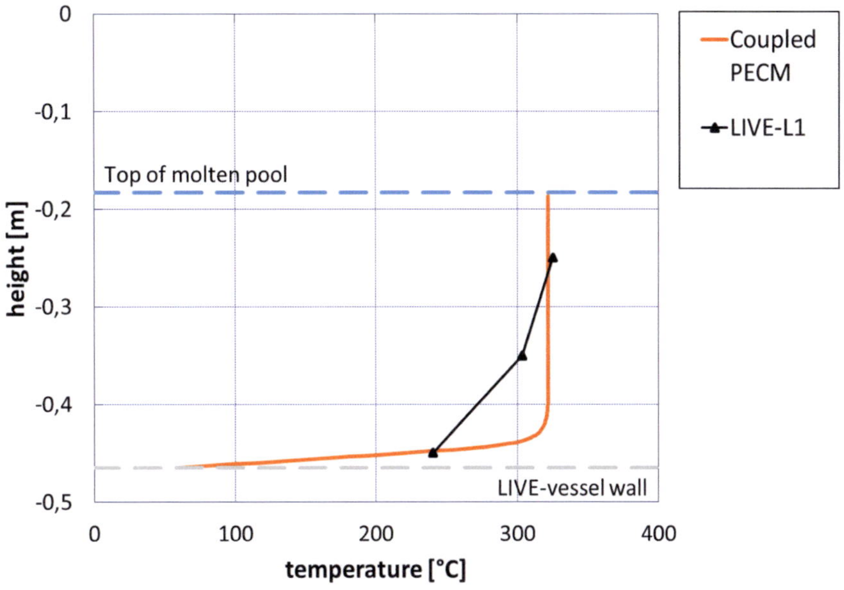

Figure 5.9: Vertical temperature profile along a vertical line at a distance of 0.175 m from the symmetry axis in the coupled *MELCOR-PECM*-simulation compared to experimental data from *LIVE*-L1

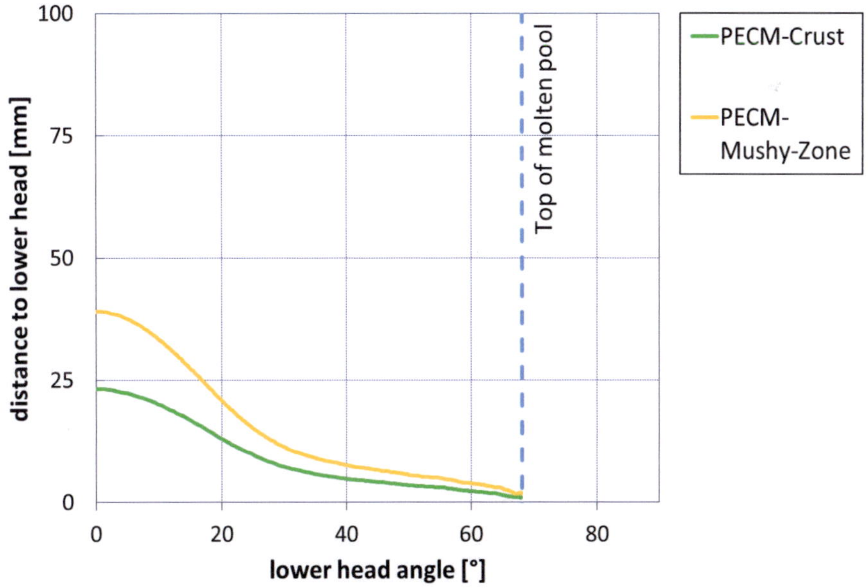

Figure 5.10: Crust and mushy-zone thickness along the *LIVE*-vessel inner wall in the coupled *MELCOR-PECM* simulation

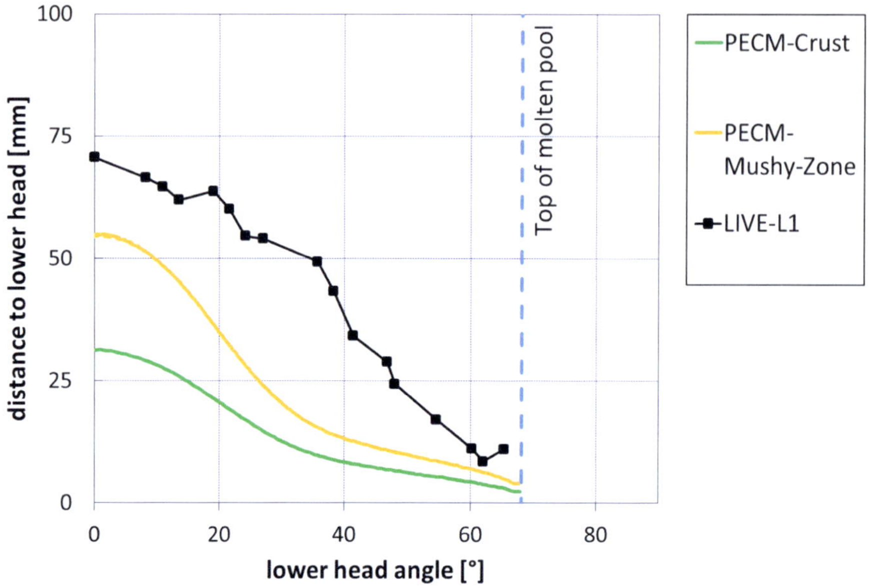

Figure 5.11: Crust and mushy-zone thickness along the *LIVE*-vessel inner wall in the coupled *MELCOR-PECM* simulation compared to experimental data from *LIVE*-L1

Based on the temperature profile shown in figure 5.7 a liquefaction indicator can be extracted. The liquefaction indicator turns to 1 if the molten material is entirely liquid and is 0 if it is solid. The distribution of this indicator for the coupled *MELCOR-PECM* system for the *LIVE*-L1 test can be seen in figure 5.12. The solid crust (blue region in figure 5.12) is thicker at the bottom than at the top of the vessel. Also, a mushy zone can be observed between the liquid and the solid material.

The simulation of the *LIVE*-L1 test with stand-alone *MELCOR* (ref. chapter 3.3) showed that the system code cannot capture the phenomena in the lower plenum in sufficient accuracy. The temperature of the molten material is overrated and a crust formation at the inner vessel wall cannot be observed. With the coupled *MELCOR-PECM* system the resolution of the temperature distribution of the molten material is possible. Despite the different temperature profiles, caused by the assumptions in the *PECM*, the mean temperature in the *PECM* matches the experimental temperature from the *LIVE*-L1 test. Furthermore, with the coupled *PECM* the formation of a crust along the *LIVE*-vessel wall can be observed well. In comparison to the experimental values there are still deviances concerning the predicted crust and mushy-zone thickness. However, the coupled simulation clearly indicates that the *PECM* improves *MELCOR*'s capabilities to simulate molten core material in the lower plenum of the *RPV* during a severe accident. In order to further test the coupled system and to expand the validation basis for *MELCOR* and the *PECM* the experiments *LIVE*-L10 and *LIVE*-L11 are simulated.

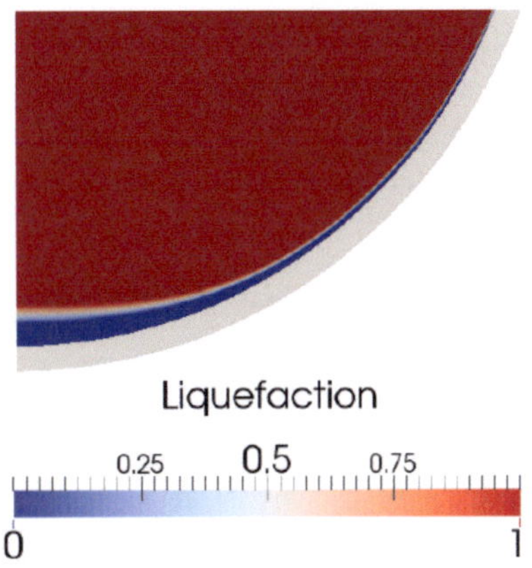

Figure 5.12: Distribution of the liquefaction indicator in *LIVE*-L1 calculated by the coupled *MELCOR-PECM* system

5.4.2 Simulation of LIVE-L10 and LIVE-L11

In the *LIVE*-L10 and *LIVE*-L11 tests the effect of different external cooling conditions was investigated. In *LIVE*-L10 a continuous support of coolant resulting in an external wall temperature below the saturation temperature of the coolant was realized. In contrast, in the *LIVE*-L11 test an adaptive coolant support was installed. The coolant support in *LIVE*-L11 is dependent on the amount of evaporated coolant resulting in a temperature of the *LIVE*-facilities outer vessel wall above the boiling temperature of the coolant. The coolant support rate in *LIVE*-L10 with 0.43 l/s is similar to the experiment *LIVE*-L1. The difference between *LIVE*-L1 and *LIVE*-L10 is the amount of molten salt inside the facility. The *LIVE*-L10 and *LIVE*-11 tests are performed with 210 l of molten salt in the facility. In order to compare the results of the two experiments the same heating power profiles were applied. Steady states were observed at heating powers of 21 kW, 16.4 kW and 9.1 kW [64]. In the following chapter the results of the steady state with a heating power of 21 kW are compared to the simulation results with the coupled *MELCOR-PECM* system. The modified calculation domain with the active surfaces for the coupling exchange and the extended grid to support the 210 l of molten material in the facility are presented in figure 5.13.

During the coupling time step the temperatures of the coolant and of the air above the molten salt in *MELCOR* are used to define the boundary conditions of the *PECM*-simulation. The temperatures of the inner vessel wall of the *LIVE*-facility are extracted from OpenFOAM and used to recalculate the corresponding temperatures in *MELCOR*. The temperature calculated by a stand-alone *MELCOR* simulation for the *LIVE*-L10 test is shown in figure 5.14.

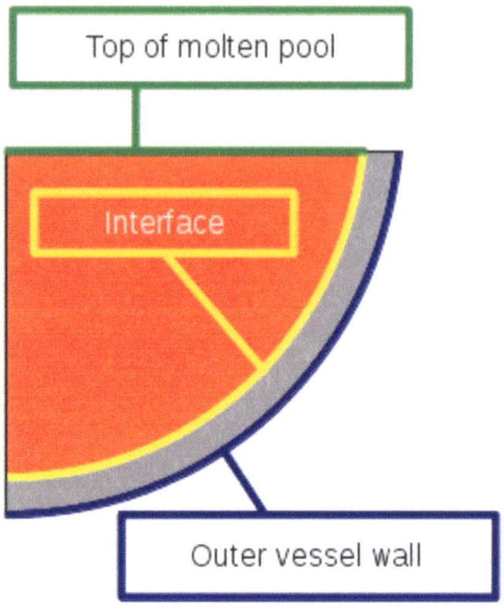

(a) Active boundaries for the coupling exchange

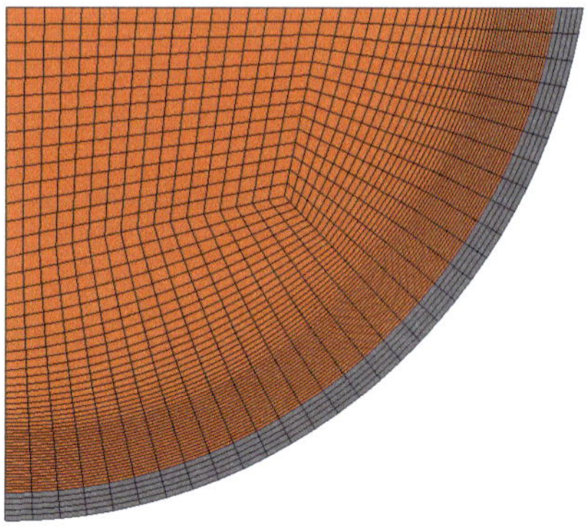

(b) Numerical grid in OpenFOAM

Figure 5.13: Calculation domain for *LIVE*-L10 and *LIVE*-L11 in the *PECM*

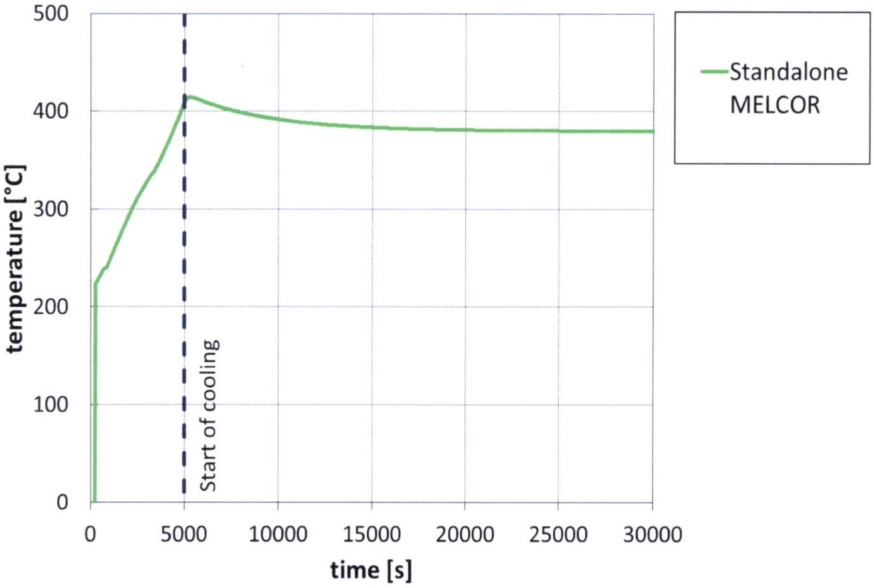

Figure 5.14: Temperature development in the molten salt pool in *MELCOR* for *LIVE*-L10

Similar to the simulation setup for *LIVE*-L1 the solid salt drops into the *LIVE*-facility at the beginning of the simulation and is then heated up volumetrically. After 5 000 seconds the entire salt is molten and the external coolant region is flooded with water. Also, the activation function in *MELCOR* activates the *PECM* in *DINAMO* (ref chapter 5.3). The *MELCOR* temperature development also shows that after about 20 000 seconds a steady state is reached. In contrast to the uniform temperature in stand-alone *MELCOR*, in the coupled simulation a temperature distribution in the molten material can be derived. The temperature distribution for the steady state of *LIVE*-L10 with a volumetric heating power of 21 kW is shown in figure 5.15.

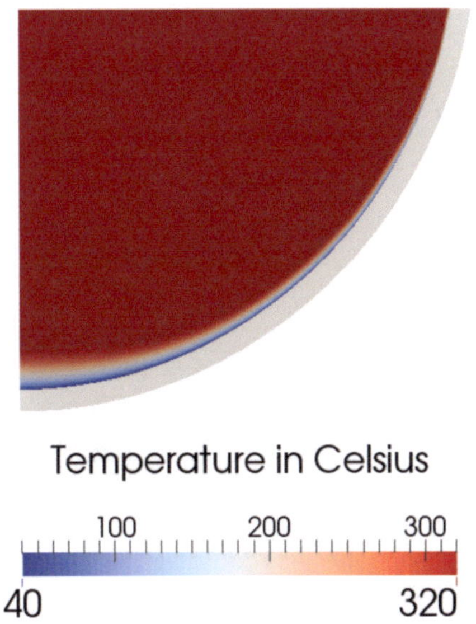

Figure 5.15: Temperature distribution in the coupled *MELCOR-PECM* simulation for *LIVE*-L10

The detailed evaluation of the *LIVE* tests L10 and L11 is carried out along a vertical line at 0.1875 m from the symmetry axis of the calculation domain. This line is located between two measurement points in the *LIVE*-facility at a radius of 0.175 m and 0.200 m from the centre of the vessel. However, none of these two measurement applications cover the entire height of the molten material. Therefore, the centre line at 0.1875 m was chosen. The arrangement of these evaluation and measurement lines in a schematic of the calculation domain is shown in figure 5.16.

The temperature profile calculated by OpenFOAM is compared to the *LIVE*-L10 experiment. This comparison is shown in figure 5.17. According to the findings in chapter 5.4.1 the *PECM* cannot reproduce the exact temperature profile measured in the experiment. Due to the assumption of a turbulent mixed pool in the entire molten material, a temperature stratification cannot be predicted. However, the mean temperature in the *PECM* is in good agreement with the mean temperature in the experiments. *MELCOR* overestimates the measured temperatures in the *LIVE* facility significantly.

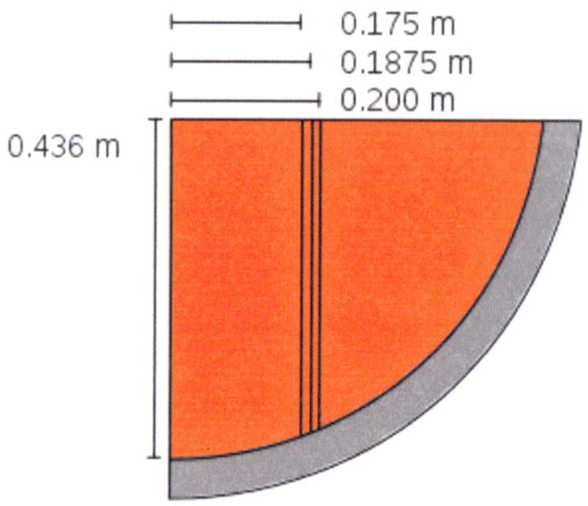

Figure 5.16: Evaluation and measurement application lines in *LIVE*-L10 and L11

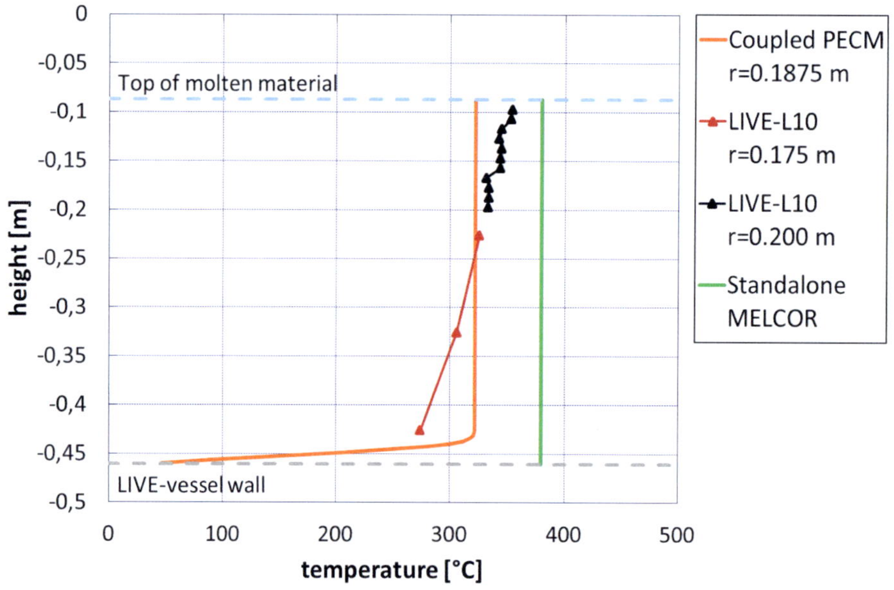

Figure 5.17: Vertical temperature profile in the coupled *MELCOR-PECM* calculation compared to stand-alone *MELCOR* and experimental results from *LIVE*-L10

A comparison of the temperatures along the inner vessel wall in stand-alone *MELCOR* and the *PECM* in the coupled simulation is shown in figure 5.18. Both *MELCOR* and the *PECM* overestimate the wall temperature in the experiment. The overall profile of the wall temperature is, however, more precisely predicted by the coupled *PECM*-simulation. The reason why the temperature profiles deviate from the simulations is the uniform coolant temperature in *MELCOR*, which is also used in the *PECM* to define the boundary condition at the outer vessel wall.

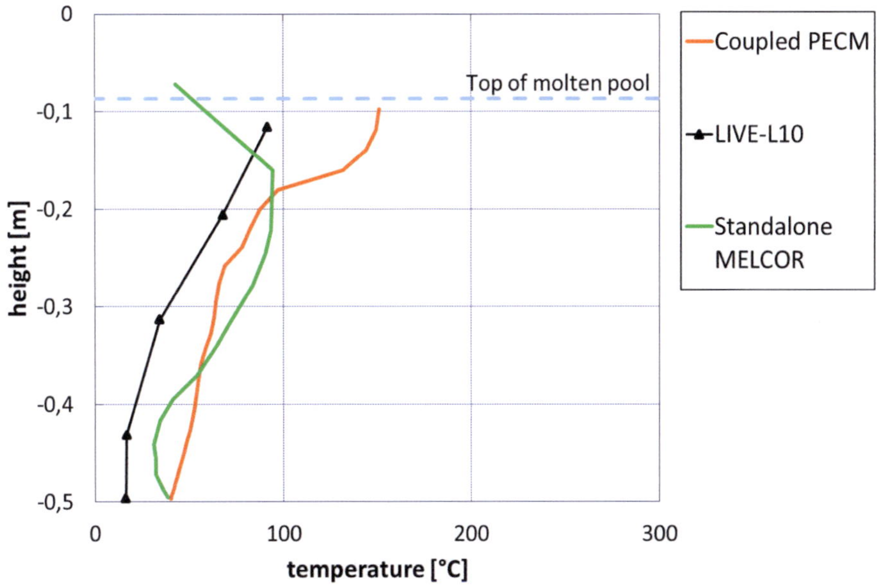

Figure 5.18: Temperatures at the inner vessel wall in the coupled *MELCOR-PECM* calculation compared to stand-alone *MELCOR* and experimental results from *LIVE*-L10

In *LIVE*-L11 test the uniform temperature at the outer vessel wall is supposed to have a minor effect on the calculation results because the heat transfer at the outer vessel wall is governed by nucleate boiling. The heat transfer coefficient for nucleate boiling is based on the saturation temperature of the coolant. The temperature development in *LIVE*-L11 calculated by stand-alone *MELCOR* can be observed in figure 5.19. According to the simulations of *LIVE*-L1 and *LIVE*-L10 the salt has to be molten before the external coolant region is flooded by water and the calculation in the *PECM* is started. Compared to the temperature calculated by stand-alone *MELCOR* for the *LIVE*-L10 test the temperature in the steady state after about 20 000 seconds is higher in the *LIVE*-L11 simulation. This is a consequence of the different heat transfer mechanism at the outer vessel wall.

In the coupled simulation, similar to *LIVE*-L10, a temperature distribution of the molten salt in the *LIVE*-facility can be obtained. The temperature profiles are evaluated according to the lines defined for the *LIVE*-L10 (ref. figure 5.16). The vertical temperature profile in the coupled simulation compared to the experimental results and a standalone *MELCOR* simulation is shown in figure 5.20. The smaller stratification gradient compared to the *LIVE*-L10 test (ref. figure 5.17) is also not resolved by the coupled *PECM*-simulation. However, with the coupled *PECM*, the mean temperature matches well with the experimental results in contrast to the stand-alone *MELCOR*. Furthermore, in *MELCOR* a higher temperature was calculated for *LIVE*-L11 than in *LIVE*-L10. In the *PECM* the mean temperatures only show slight deviances. This is in accordance with the experiments.

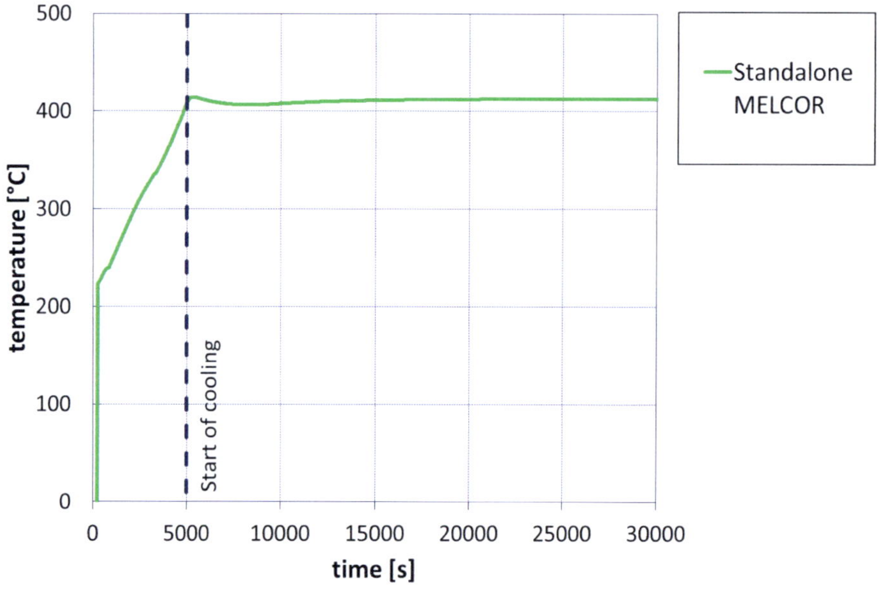

Figure 5.19: Temperature development in the molten salt pool in *MELCOR* for *LIVE*-L11

The temperatures at the inner wall surface of the *LIVE*-vessel are shown in figure 5.21. The temperatures in the coupled *MELCOR-PECM* simulation match the temperatures found in the experiment very well. In contrast, in the stand-alone *MELCOR* simulation the temperature in the middle of the profile is overestimated. The overall profile as well as the exact temperature data is reproduced very well by the coupled simulation with the *PECM*.

As already observed in the simulations on the *LIVE*-L1 test, the *PECM* coupled to *MELCOR* has the capability to improve *MELCOR*'s prediction of molten core material in the lower plenum during a severe accident in an *NPP*. In the coupled *PECM* the mean temperature of the simulated *LIVE*-tests are predicted very well. However the assumption in the *PECM* that the entire pool is governed by the characteristic velocities, leads to deviations in the temperature profile. The temperature stratification, which was found in the *LIVE*-experiments, could not be

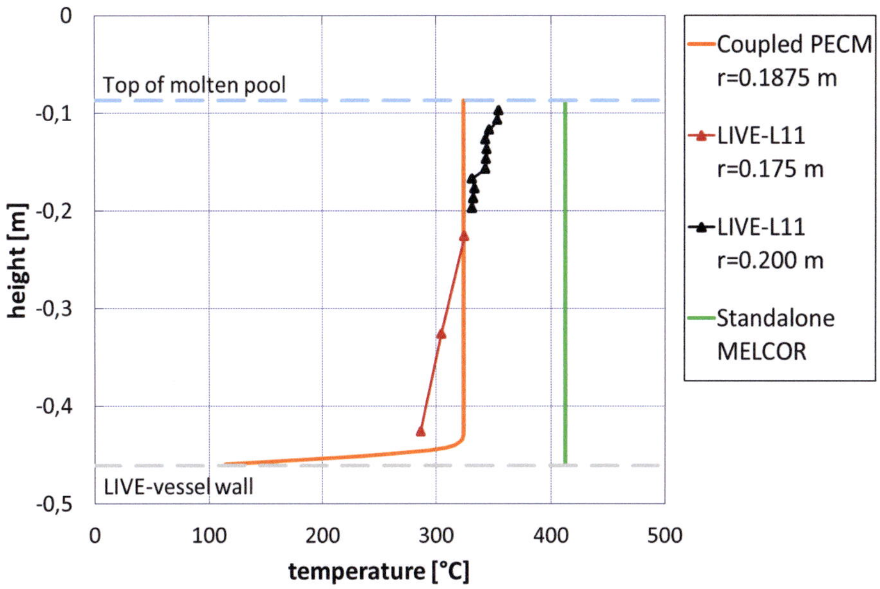

Figure 5.20: Vertical temperature profile in the coupled *MELCOR-PECM* calculation compared to stand-alone *MELCOR* and experimental results from *LIVE*-L11

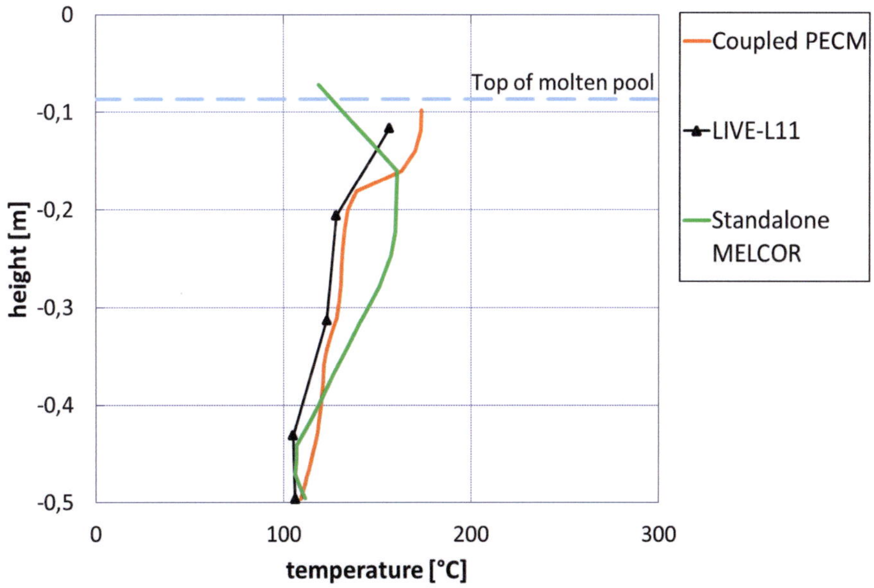

Figure 5.21: Temperatures at the inner vessel wall in the coupled *MELCOR-PECM* calculation compared to stand-alone *MELCOR* and experimental results from *LIVE*-L11

reproduced. The temperatures at the inner lower head surface, which significantly influence the heat flux through the lower head, show deviations when the heat transfer at the outer lower head surface is governed by conduction. Regarding nucleate boiling at the outer vessel wall, the temperatures at the inner vessel surface show a very good agreement with the experimental findings. The coupled *PECM* improves the prediction of the temperatures better than the stand-alone *MELCOR* calculation. The simulations of *LIVE*-L10 and *LIVE*-L11 showed that *MELCOR* significantly overestimates the temperature in the molten salt and is unable to reproduce the temperature profile along the inner vessel wall. Using the coupled *PECM* these deviances of the stand-alone *MELCOR* simulation are addressed and the simulation accuracy is improved significantly.

6 Coupled simulation of a nuclear power plant

The simulations performed on the different *LIVE*-tests showed the ability of the coupled *MELCOR-PECM* system to receive a more detailed description of the molten core material in the lower plenum. In this chapter an entire Nuclear Power Plant (*NPP*) will be simulated under severe accident conditions to show the applicability of the coupled systems to traditional *MELCOR* simulations. Similar to the previous chapter in these coupled simulations the original Phase-Change Effective Convectivity Model (*PECM*) approach shall be used due to the cooling conditions of the molten core material in the lower plenum. Prior to the coupled simulations an input for a generic power plant will be developed in *MELCOR*. This input deck will be used to perform a coupled simulation of a severe accident with the In-Vessel Melt Retention (*IVR*) (ref. chapter 2.2) with the coupled *MELCOR-PECM* approach.

6.1 Input deck for a generic pressurized water reactor

The *MELCOR* input for the generic Pressurized Water Reactor (*PWR*) has a thermal power of 3 850 MW and shows four steam generator loops. In the postulated accident scenario a leak of 50 cm^2 (Flow Path (*FL*) 228 in figure 6.1) opens in the cold leg of the loop, which also contains the pressurizer. The high and low pressure injection pumps are not available which leads to a meltdown of the core components due to the decay heat. After the Emergency Shutdown of the Nuclear Reactor (*SCRAM*) is initiated the reactor cavity is flooded with water. The coolant support of the cavity is dependent on the water level. It ensures a sufficient coolant support to keep the outer boundary of the lower head permanently covered with water. In figure 6.1 the thermohydraulic nodalisation of the input deck is shown. In this figure the steam generator loop with the pressurizer is illustrated. The other loops are nodialised in a similar way.

The core region represented by the Control Volumes (*CV*s) 102 and 110 is divided into 6 rings with 14 axial levels in the COR-package. The total height of the core region is 5.35 m. The 2 upper and lower levels of the core region (grey) represent the upper plug, the lower plug and the lower core support plate. The active core region (orange) has a height of 3.9 m and is divided into 10 axial levels. The core is formed by 57 300 fuel rods and 1 464 control rods made of AgInCd. The outer core ring (6) also contains the core shroud as well as the bypass volume (light blue). A schematic of the nodalisation in the COR-package of *MELCOR* is illustrated in figure 6.2.

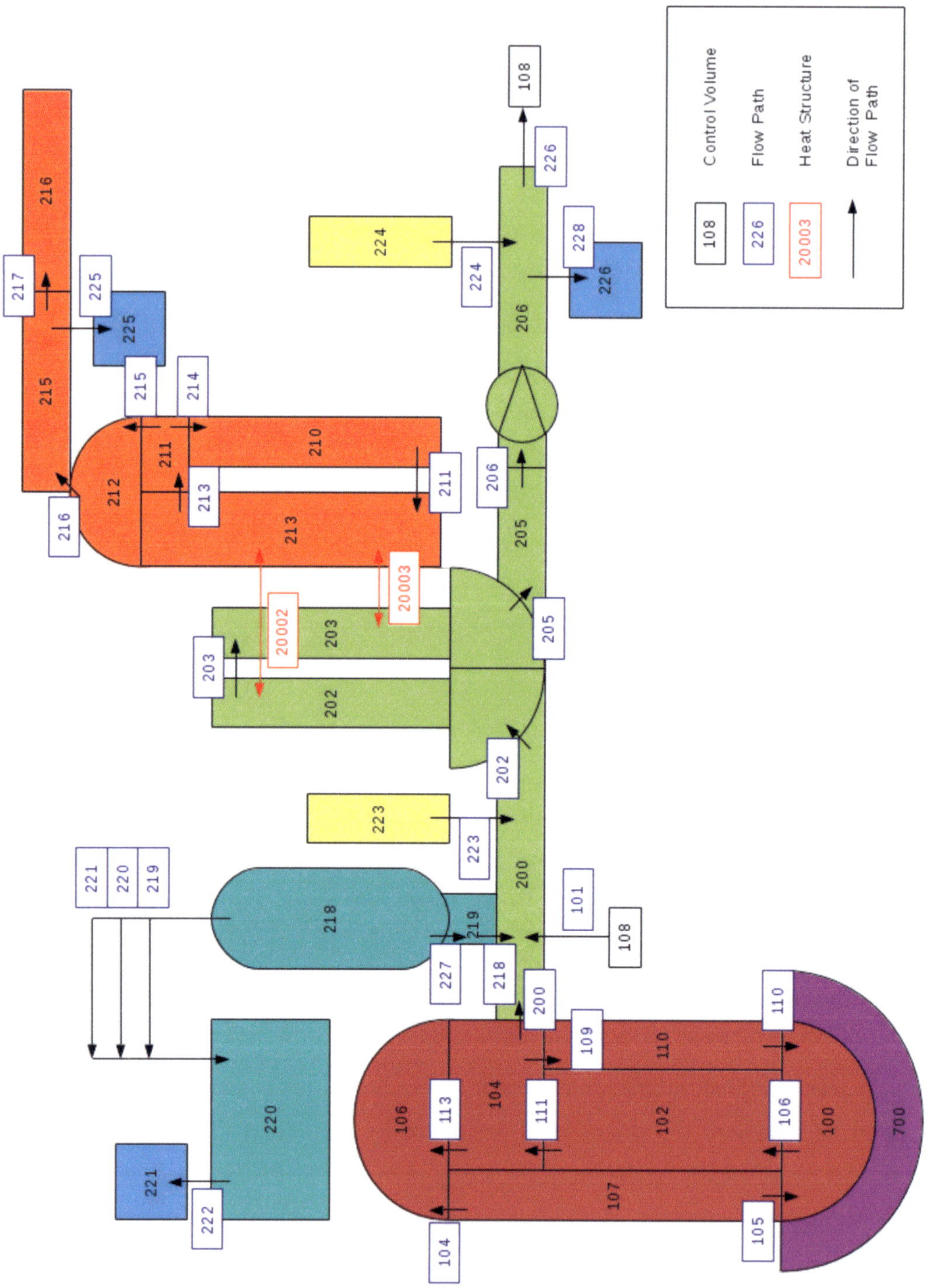

Figure 6.1: Thermohydraulic nodalisation scheme of the generic *NPP* used to test the coupled *MELCOR-PECM* system

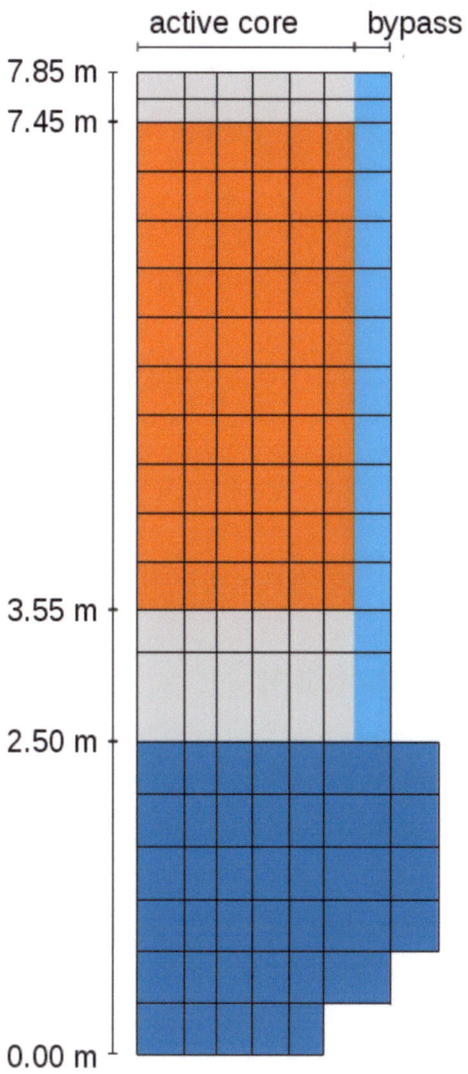

Figure 6.2: Nodalisation in the COR-package of the generic *NPP* used to test the coupled *MELCOR_PECM* system

The lower plenum is represented by the dark blue region in the nodalisation in the COR-package (ref. figure 6.2). An additional ring is introduced to capture the Reactor Pressure Vessel (*RPV*) with a radius of 2.5 m. 6 axial levels and 7 radial rings form the lower plenum, which is represented by *CV* 100 in the thermohydraulic scheme (ref. figure 6.1). The hemispherical lower head of the *RPV* with a radius of 2.5 m is divided into 12 sections with 6 layers each according to the nodalisation of the lower plenum in the COR-package. An illustration of the positions of these segments along the curved lower head is shown in figure 6.3.

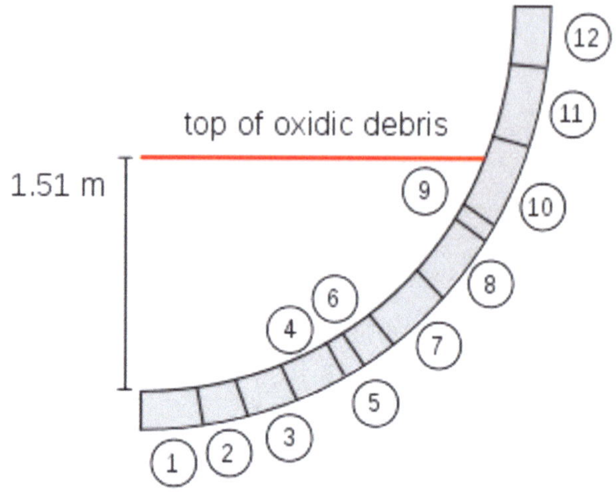

Figure 6.3: Nodalisation of the lower head of the generic *NPP* used to test the coupled *MELCOR-PECM* system

6.2 Coupled simulation of a severe accident

The results of the coupled simulation are compared to a standalone *MELCOR* calculation of the same severe accident. The degradation of the core is initiated by loss of coolant through the leak represented by *FL* 228 (ref. figure 6.1). The main coolant pump (marked in *CV* 206 in figure 6.1) is shut down due to the increased void of the coolant in the primary circuit. The remaining coolant left in the primary circuit is not enough to cool the core, which is heated by the decay heat. The evolution of the decay heat is provided by a table in *MELCOR*. The mass flow rate of water and steam through the leak is shown in figure 6.4.

As shown in figure 6.5 at first the metal components of the core melt and relocate to the lower plenum. After 8 000 seconds the lower core support structure fails and a degraded core material relocates to the lower plenum. At 15 000 seconds the amount of debris in the lower plenum has reached an almost steady state, which is suitable for simulations with the *PECM*. The mass of oxidic melt, which is described by the *PECM*, is approximately 153 tons. Assuming an oxidic layer containing the degraded fuel and the oxidised metals, 123 tons of oxidic debris accumulates in a pool height of 1.51 m as indicated in figure 6.3.

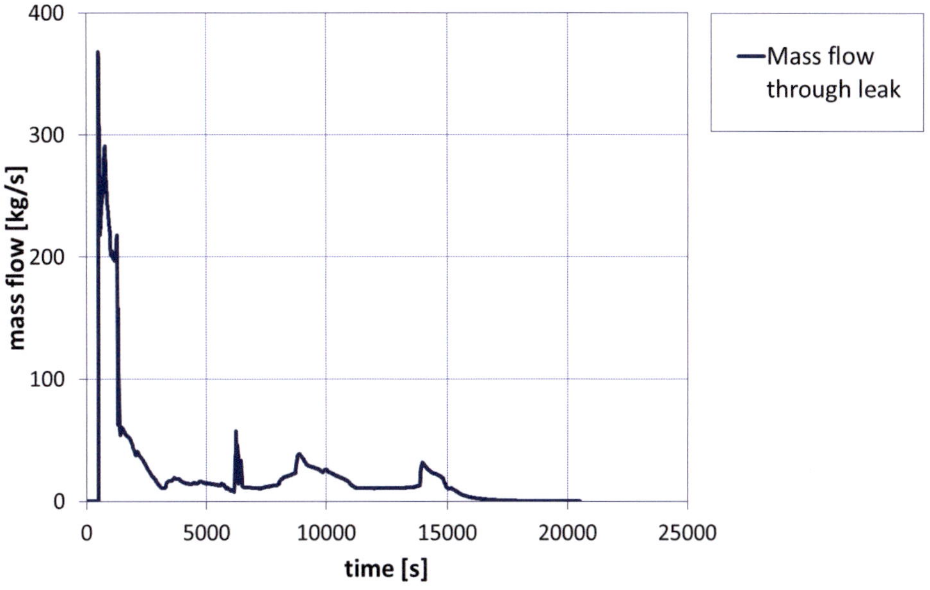

Figure 6.4: Mass flow rate through the leak in the cold leg *FL* 228

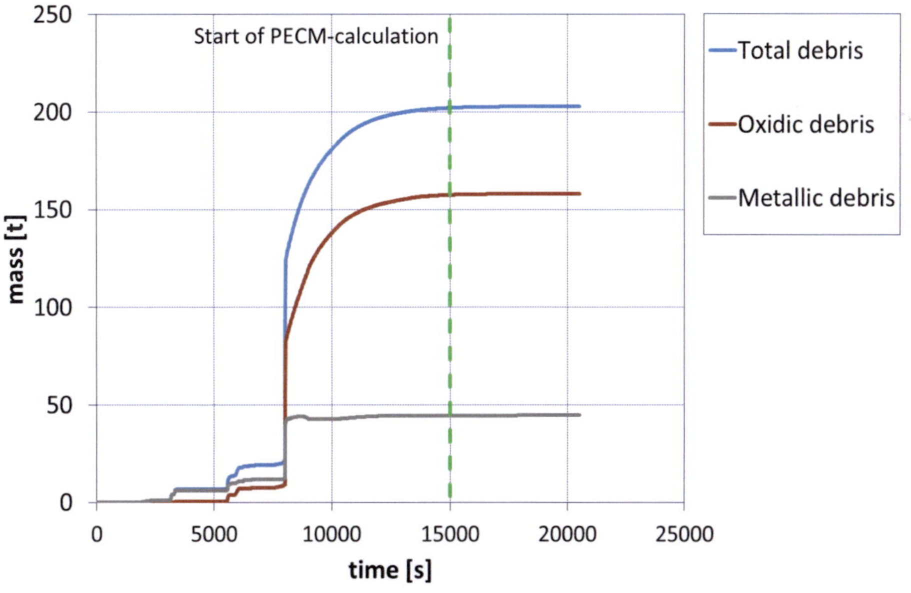

Figure 6.5: Debris masses in the lower plenum

In the standalone *MELCOR* simulation a melt pool is predicted in the lower plenum with a temperature of about 2 400 °C. The mass of this molten material is 7.5 tons. This corresponds to 6.1 % of the entire mass of oxidic material in the lower plenum. The temperature in the molten material as well as the evolution of the mass of the pool in the lower plenum are shown in figure 6.6.

Due to the amount of debris in the lower plenum and the heat flux through the lower head, the lower head fails in segment 10 (ref. figure 6.3) after 20 514 seconds and debris is ejected into the cavity. As soon as the failure is detected the calculation is terminated.

In the coupled simulations the coupling methodology was modified slightly to match the different requirements of the entire power plant. The temperature of the coolant in the reactor cavity is replaced by the temperatures along the outer surface of the lower head in *MELCOR*. As the upper boundary of the molten material the temperature of the metal above the oxidic material is used. The calculation domain for the *PECM* in OpenFOAM with the active surfaces for the coupling as well as the numerical grid is presented in figure 6.7.

Figure 6.8 shows the development of the *RPV*-pressure in the coupled simulation compared to the standalone *MELCOR* simulation. Also, the water level in the *RPV* is shown. After 500 seconds of normal operation a leak of 50 cm^2 in the cold leg opens. This results in an abrupt decrease of the pressure and water level in the *RPV*. The relocation of core material to lower parts of the reactor leads to a temporary increase of the water level after about 6 000 seconds. The decay heat evaporates the water in the *RPV*, which results in the monitored temporary increased pressure. Until the end of the coupled calculation the parameters of the two simulations show the same behaviour.

The most significant difference between the calculation results of the coupled simulation and standalone *MELCOR*, is the failure time of the lower head. As indicated in figure 6.8 in the coupled simulation the lower head fails approximately 1 132 seconds after the start of the external calculations in the *PECM* at 15 632 seconds after the opening of the leak. In standalone *MELCOR* the failure is predicted about 20 000 seconds after the leak opened. The coupled simulation predicts the collapse almost 1 hour and 15 minutes sooner, which is a significant amount of time with regard to the accident progression and the possible initiation of safety measures in an actual severe accident. The failure in both the coupled and the standalone simulation is induced by the abrupt increase of the temperature in the lower head when the Critical Heat Flux (*CHF*) is exceeded (ref. Cheng and Müller [137] or Hall and Mudawar [138]). The temperature development at the outer surface of the lower head is shown in figure 6.9 for the coupled and for the standalone simulation.

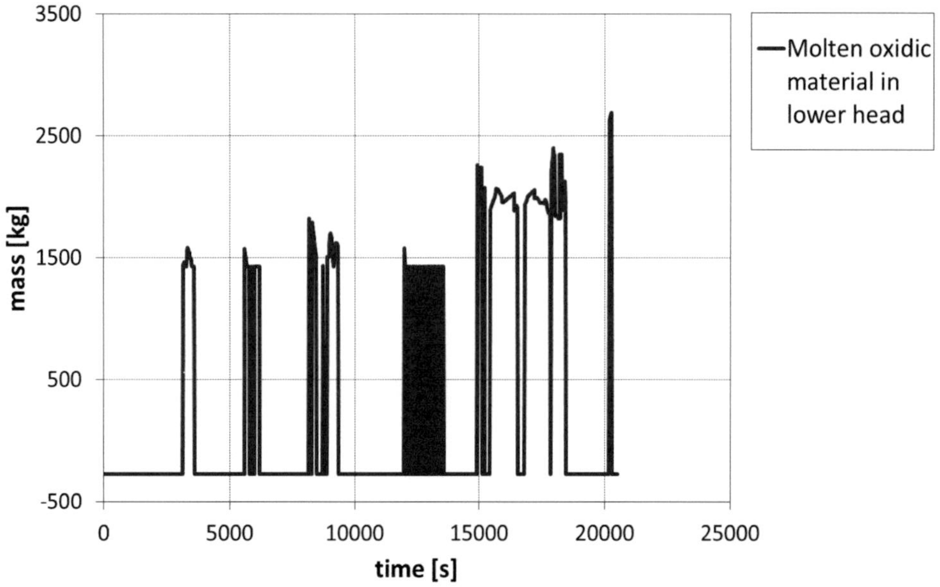

(a) Temperature evolution

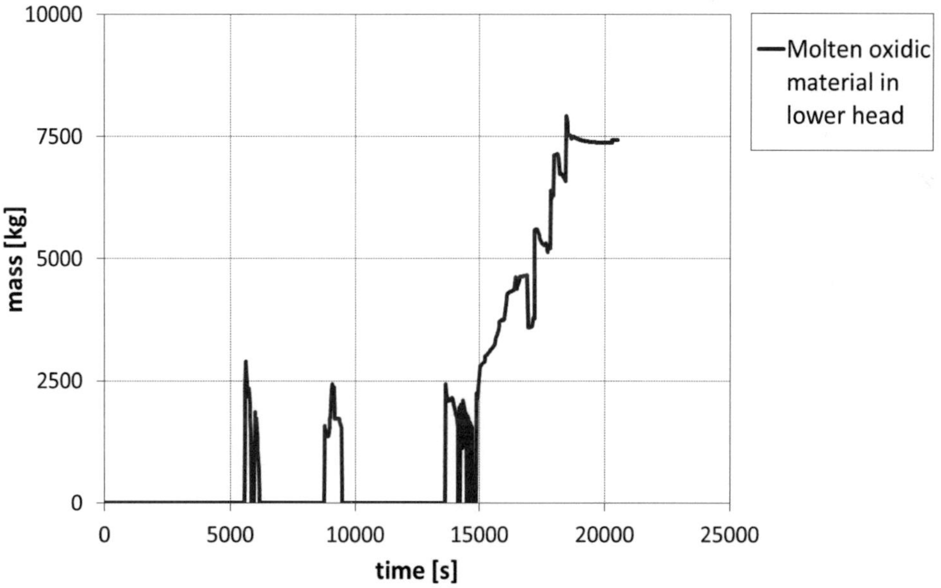

(b) Mass of molten material

Figure 6.6: Temperature and mass of the molten oxidic material in the lower plenum of the *RPV*

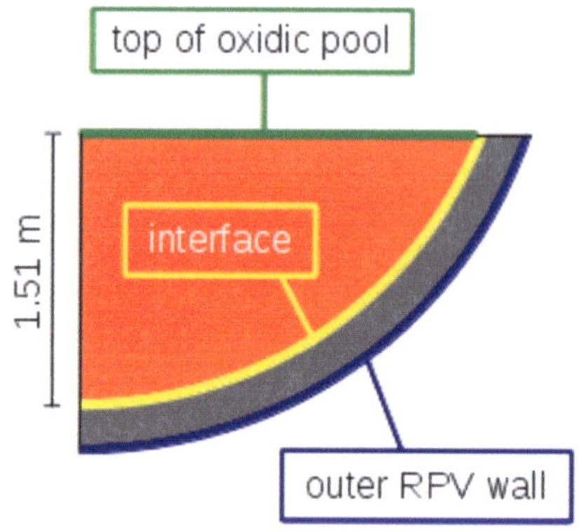

(a) Active boundaries in the coupling exchange

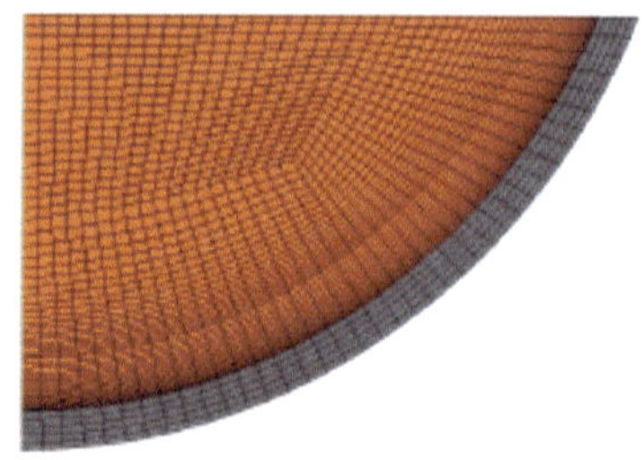

(b) Numerical grid for the *PECM*

Figure 6.7: Calculation domain in OpenFOAM for the coupled simulation

In the standalone *MELCOR* simulation the temperature in segment 10 shows an abrupt temperature rise, which leads to the failure of the *RPV*. In the coupled simulation segment 10 shows the sharp temperature rise, which is also monitored in segment 8. According to the *MELCOR* output file for the coupled simulation the lower head in segment 8 fails first and causes the simulation to stop. The outer wall temperature is dependent on the current heat flux, which in turn dependeds on the temperature at the inner surface of the lower head. The development of these temperatures at the interface between the core material and the steel structure of the lower head is shown in figure 6.10.

According to the coupled *PECM* the temperature in the lower head rises faster compared to the standalone *MELCOR* calculation leading to a higher temperature at the inner surface. This causes a higher heat flux through the *RPV* wall leading to the failure of the lower head when exceeding the *CHF*. In comparison to the standalone *MELCOR* simulation, the coupled calculation allows for the observation of the temperature distribution in the molten material and the lower plenum. The temperature of the oxidic debris, calculated with the *PECM*, at specific simulation times in *MELCOR* is shown in figure 6.11.

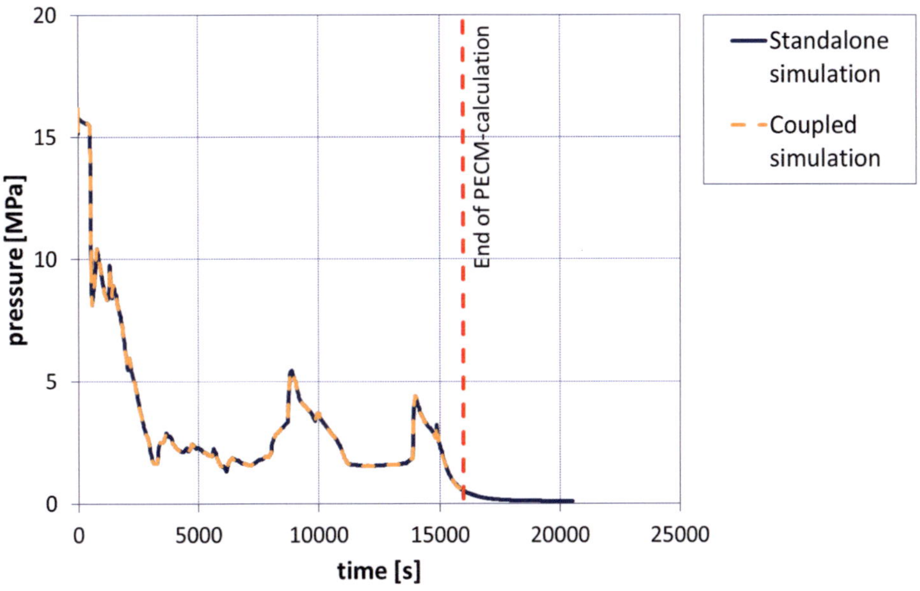

(a) *RPV*-pressure

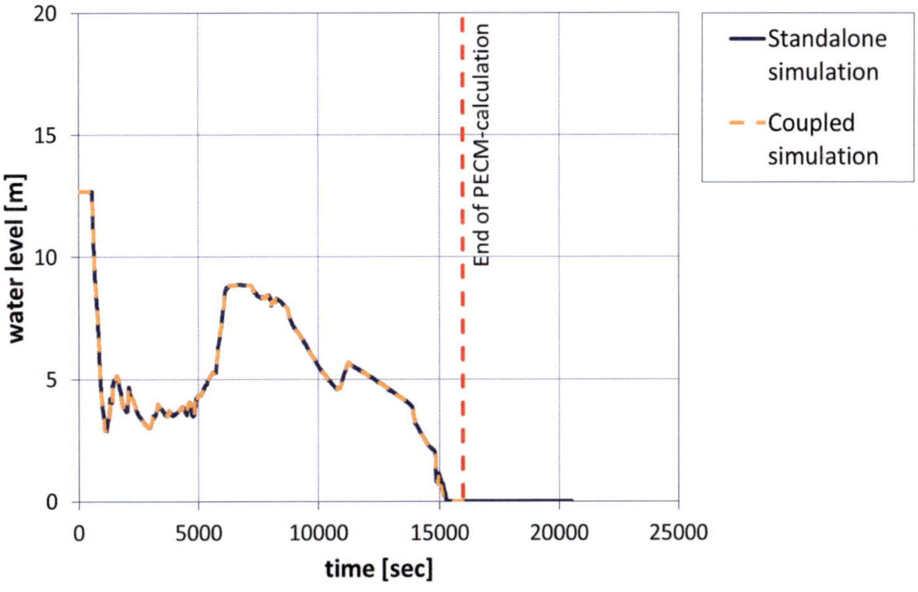

(b) Water level in *RPV*

Figure 6.8: Pressure and water level development in the coupled severe accident simulation compared to standalone *MELCOR*

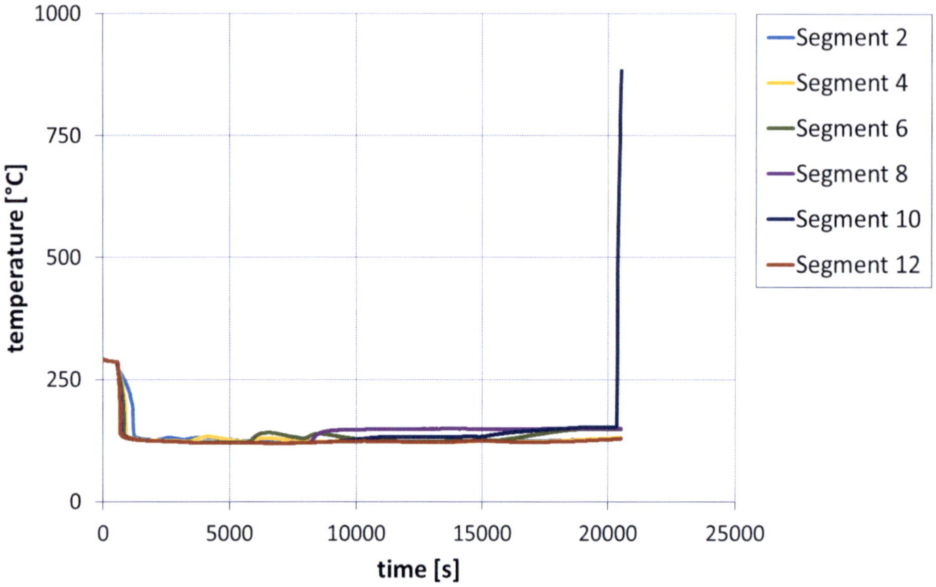

(a) Standalone simulation

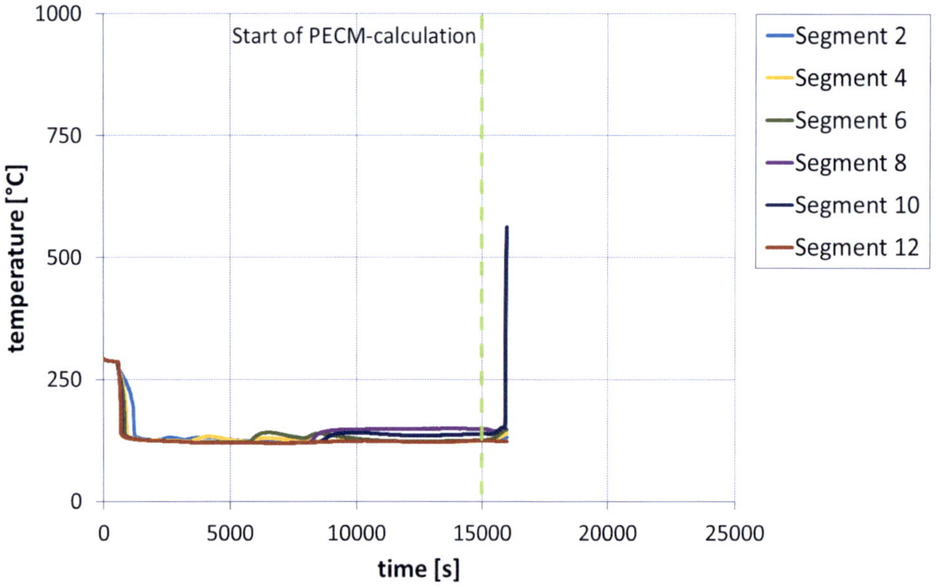

(b) Coupled simulation

Figure 6.9: Temperatures of the outer lower head surface

The temperature distribution at the beginning of the simulation in the *PECM* shows diffusive heat transfer close to the lower head. As the simulation progresses this solid material melts and is treated by the *PECM* as liquid material. In contrast to the simulation results in standalone *MELCOR*, where only about 6 % of the oxidic material is liquefied, in the coupled simulation almost all of the oxidic material in the lower plenum is molten. Furthermore, a higher temperature of the molten pool is predicted by the *PECM*. In figure 6.11 the increse of the lower head temperature due to the hot molten core materiral can be observed. The significant higher amount of hot molten material close to the lower head caused the sharp temperature increase shown on the right side of figure 6.10.

The simulation of an entire *NPP* during a severe accident with the coupled *MELCOR-PECM* system has shown the applicability of this system for further severe accident analyses. According to the coupled system the behaviour of the molten oxidic pool in the lower plenum was resolved in more detail. The results of the coupled simulation compared to the standalone *MELCOR* simulation showed a significant deviation in the scenario of lower head failure. In the coupled calculation the lower head failed approximately 1 hour and 15 minutes earlier than predicted by standalone *MELCOR*. Taking into account the total duration since the initiation of the severe accident of about 5 hours this deviation is essential in the prediction of the accident progression and the applicability of Severe Accident Management (*SAM*) strategies. Based on the more accurate prediction of various *LIVE*-tests, the simulation result of the coupled simulation can be considered to be more precise and realistic. This first coupled simulation to analyse the *IVR*-strategy on an *NPP* can predict the temperature distribution and development in the molten material in more detail compared to the standalone *MELCOR* calculation.

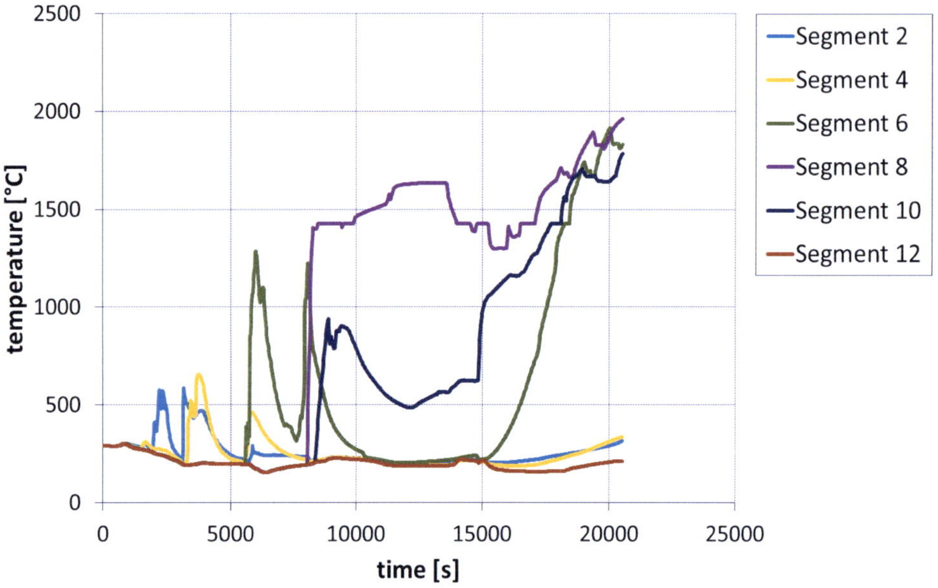

(a) Standalone simulation

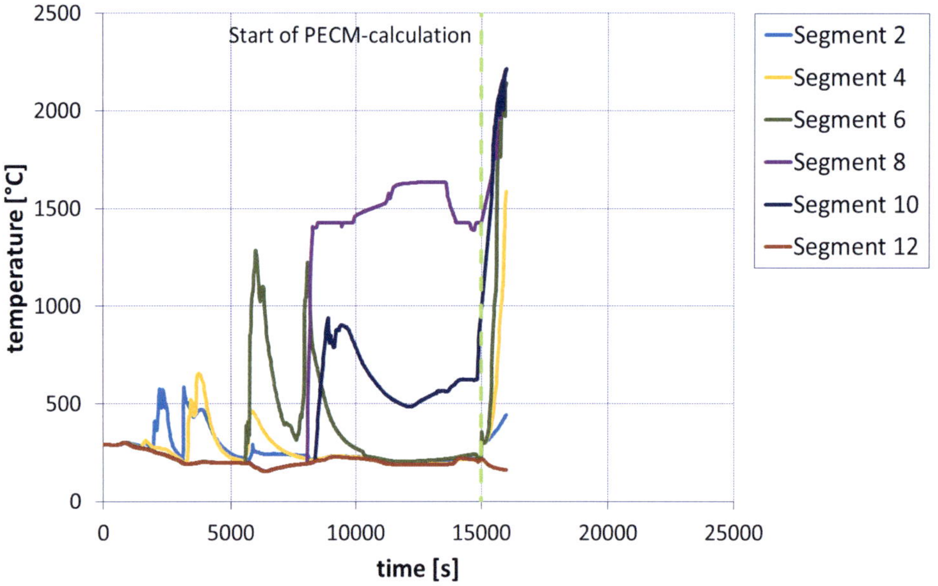

(b) Coupled simulation

Figure 6.10: Temperatures of the inner lower head surface

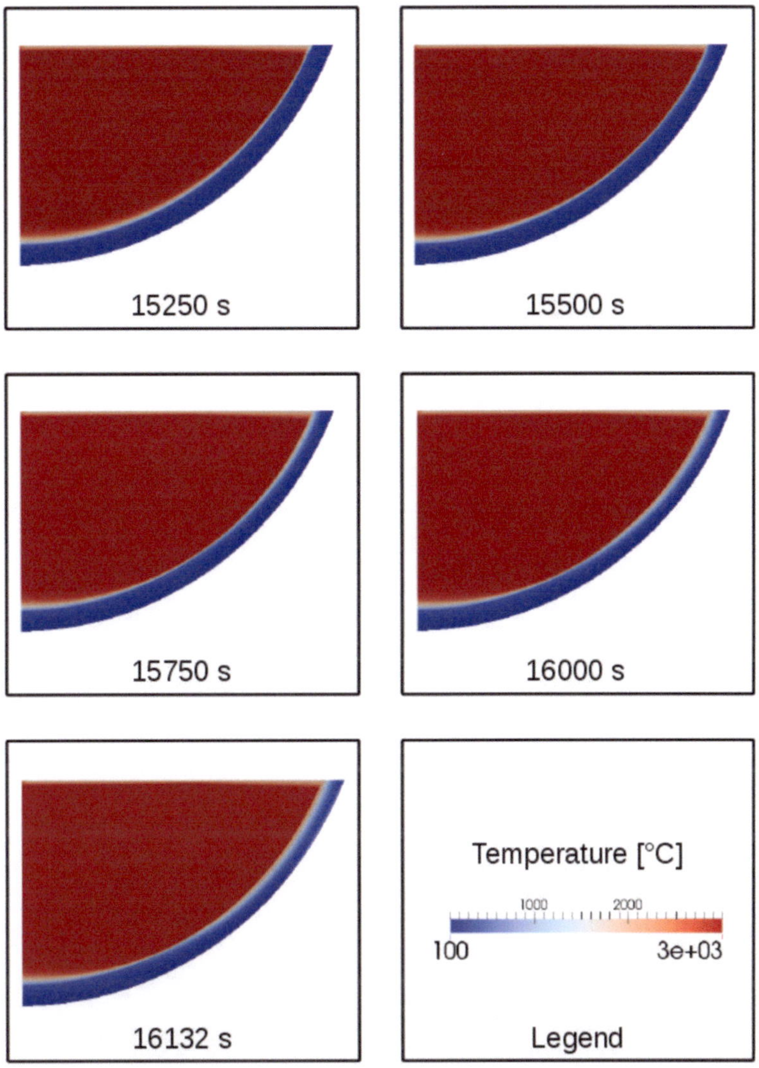

Figure 6.11: Temperature distribution in the oxidic material and the adjacent part of the lower head calculated by the *PECM* in a coupled simulation

7 Summary and Conclusion

In the presented work the In-Vessel Melt Retention (*IVR*) accident management strategy was in focus. This strategy aims at the retention of molten core material inside the Reactor Pressure Vessel (*RPV*) in order to preserve its integrity. Core material can melt and relocate to the lower plenum in the progression of a severe accident, which can be caused by a Loss of Coolant Accident (*LOCA*). The *IVR*-strategy has the ability to stop the progression of a severe accident inside the *RPV* and prevent radioactive material from entering the containment building or the environment. Several experimental studies have been performed to investigate external cooling of the *RPV* when there is an internally heated melt inside the lower plenum. Besides the experimental investigation of the accident management strategy numerical investigations have been conducted as well. A model to describe the behaviour of molten core material in the lower plenum is the Phase-Change Effective Convectivity Model (*PECM*), which was developed at the Royal Institute of Technology (*KTH*).

The progression of a hypothetical severe accident in a Nuclear Power Plant (*NPP*) can be predicted using specialized integral codes such as *MELCOR* or *ATHLET-CD*. These codes are able to simulate the entire *NPP*. The integral approach used in system codes includes several assumptions and simplifications to ensure fast and stable code performance. For a detailed analysis of specific phenomena, such as the behaviour of molten core material in the lower head, these integral codes often cannot provide a sufficiently detailed result. Approaches such as Computational Fluid Dynamics (*CFD*)-codes can resolve the physical processes in detail, however, the computational effort for such simulations is very high. For an entire *NPP* a detailed simulation is not suitable due to the necessary amount of computational resources and the calculation time. The coupling of new models or entire external programs with integral codes allows for detailed simulations in the context of a severe accident. The coupled special code or model (e.g. *PECM*) uses the boundary conditions provided by the integral code (e.g. *MELCOR*), which simulates the entire nuclear power plant, to simulate a specific phenomena. This coupled approach allows for the simulation of an entire severe accident in a reasonable amount of calculation time using detailed codes for specific phenomena.

In this work the *LIVE*-facility was calculated by the integral code *MELCOR*. Therefore, an input for this code was developed based on the geometric and physical properties of this experimental facility. This input was tested by simulating the *LIVE*-I1 test, which uses water as simulant material for the molten core material. The calculation results showed that in *MELCOR* the heat transfer from the lower head to the adjacent external coolant is not considered if the temperature of the lower head is below the boiling temperature of the coolant. For almost all *LIVE*-tests this subcooled external condition was applied. In order to simulate the experiments performed in the *LIVE*-facility with *MELCOR*, a heat transfer correlation for subcooled water was implemented into the *MELCOR* source code.

The simulation of the experiment *LIVE*-L1 showed that *MELCOR* does not capture the behaviour of the molten pool material in the lower plenum in sufficient detail. In *MELCOR* the temperature of the molten material is assumed to be uniform for the entire pool. Further, the crust fromation along the lower head walls, which was monitored in the experiment, is not predicted. Simulations perfomed on *LIVE*-L1 with the *PECM*, which was implemented into the *CFD*-software OpenFOAM, showed a significantly better resolution of the material in the lower plenum. The mean temperature in the experiment is captured by the *PECM*-simulation very well. Furthermore, the *PECM* is capable to resolve the temperature distribution in the molten pool and to predict the formation of a crust along the cooled *LIVE*-vessel wall. These simulations indicate that the *PECM* can improve the prediction of molten core material in the lower plenum during a severe accident in *MELCOR* significantly.

The capability of the *PECM* to provide a detailed simulation of the behaviour of molten core material in the lower plenum was tested on additional coolant strategies to validate this model further. In contrast to the experiment *LIVE*-L1, in the *LIVE*-L7v test in addition to the external cooling throught the *LIVE*-vessel wall, the melt was cooled from the top. A specially designed cooling lid was able to directly cool the upper boundary of the molten material. The simulation results showed that despite the good match of the mean temperatures in the previous simulations the temperature calculated for different steady states in *LIVE*-L7vsignificant differences from the experimental findings can be found. Therefore, modifications were made to the *PECM* methodology. According to the theory of an internally heated fluid, which is cooled from the top and through the side walls, the pool is divided into a stratified and a turbulent mixed regime. In the *PECM* the pool is treated as one regime governed by turbulent mixing. In order to determine the optimum height for the subdivision of the different regimes a parameter study was performed. In the stratified part an effective conductivity approach is used. Here the conduction in the horizontal direction is multiplied with the local sideward Nusselt-number resulting in a stratified temperature profile in this region. In the upper region, governed by turbulent mixing, the traditional *PECM*-approach is used. For the different steady states in *LIVE*-L7v the determined factor is dependent on the geometric properties of the *LIVE*-vessel and can be updated automatically at every time step of the calculation. With the Stratified Approach for the Phase-Change Effective Convectivity Model (*S-PECM*) the temperatures measured in the experiment were predicted very well. Also a stratified region was identified in the lower part of the *LIVE*-vessel. Still some differences in the stratified region and a slight difference in the local temperature distribution could be observed compared to the experimental findings. In contrast to the traditional *PECM*-approach, a significantly better agreement with the *LIVE*-L7v test could be monitored. This was confirmed by the comparison of crusts thicknesses in the *PECM* and *S-PECM* with the data from the experiment.

In order to couple the *PECM* or other auxillary models to the integral code *MELCOR* the external tool *DINAMO* was developed. This program contains a coupling interface to communicate with the integral code *MELCOR* via the program MPIEXEC. Furthermore, *DINAMO* contains an interface to address new and auxillary models, which are intended to be coupled to *MELCOR*. Using *DINAMO* the development of a special coupling interface for each model or program becomes obsolete and the otherwise complex modification of the *MELCOR* source code can be avoided.

The coupling of new models to *MELCOR* using *DINAMO* can be classified as external, explicit and synchronously sequential. The external and explicit character is due to the definition of the coupling interface in *MELCOR* and MPIEXEC. Traditionally an asynchronous coupling exchange, in which both the coupled program and *MELCOR* pursue their own time step progress between two predefined coupling time steps, is implemented in the coupling interface. Investigations performed within this work have demonstrated that this asynchronous coupling methodology significantly effects the calculation results compared to a standalone *MELCOR* simulation. The asynchronous coupling changed the *MELCOR* system timestep, which influenced the simulation results significantly. As a consequence a synchronized coupling methodology between *MELCOR* and *DINAMO* was developed and tested by the coupling of the Larson-Miller-Model to simulate the failure of the lower core support plate during a severe accident. The sequential character of the coupled models in *DINAMO* is similar to the implementation and communication of the different packages in the program *MELCOR*.

By coupling an additional or new model to the system code *MELCOR* not only the calculation accuracy in the system code is improved. Moreover, the dynamic boundary conditions in the external models defined by the simulation results in the system code additionally allow for a direct adaptation to the transient changes in the surrounding system. In order to expand the validation basis for the *PECM* and to demonstrate the benefits of the coupled system simulations of experiments conducted in the *LIVE*-facility at the Karlsruhe Institute of Technology (*KIT*) were performed. In the *LIVE*-facility the behaviour of internally heated liquid material in the lower plenum of a *RPV* is investigated.

The *PECM* in OpenFOAM was coupled to the system code *MELCOR*. The interface in *DINAMO* was modified to control the execution of the *PECM*-solver in OpenFOAM, which was developed based on the original User Defined Function (*UDF*) for ANSYS FLUENT used at the *KTH*. The boundary conditions of the simulation setup in OpenFOAM can also be modified by the interface in *DINAMO*. In the coupled calculation the boundary conditions for the calculations in the *PECM* are provided via the coupling interface from the simulation of the entire facility (*LIVE*-facility or *NPP*) in *MELCOR*. At the end of the simulation timestep in OpenFOAM the temperature distribution along the interface between the molten material and the adjiacent wall is extracted from the *PECM* calculation and sent via the coupling interface to *MELCOR*. This data is used to recalculate the temperatures in the lower head.

The coupled *MELCOR-PECM* system was tested by the simulation of the experiment *LIVE*-L1. The coupled simulations were performed with the original *PECM* due to the good agreement with the experimental findings for this specific cooling configuration. In the coupled simulation the mean temperature measured in the experiment was reproduced accurately. In the coupled simulation also additional information on the behaviour of the molten material, such as the temperture distribution in the pool or the formation of the crust along the cooled lower head, could be obtained. This simulation showed the applicability of the *MELCOR-PECM* system to calculate experiments in the *LIVE*-facility. Furthermore, the capabilities of *MELCOR* to simulate molten core material in the lower plenum of the *RPV* were expanded. Transient boundary conditions based on the *MELCOR*-simulation were applied to the *PECM*-simulation in OpenFOAM.

Further validation of the coupled system and the *PECM* has been carried out by simulations of the *LIVE*-tests L10 and L11. These tests were inspired by the results of this work and investigate the influence of the external cooling condition on the melt pool behaviour. In *LIVE*-L10 similar to *LIVE*-L1 the temperature of the coolant, which is below the boiling temperature, is realized by a constant water support. In contrast in the *LIVE*-L11 test the coolant support depends on the amount of steam leaving the coolant area caused by nucleate boiling at the outer vessel wall. The two experiments were also simulated with standalone *MELCOR* to show the improvements of the coupled approach. For *LIVE*-L10 the coupled simulation matches the mean temperature in the facility very well, yet a temperature stratification cannot be predicted. The standalone *MELCOR* calculation significantly overestimates the temperature of the molten material. Both the coupled simulation and the standalone simulation overestimate the temperature at the inner lower head surface, yet the profile is captured better by the coupled simulation. The simulation of *LIVE*-L11 also shows a good agreement of the mean temperature of the coupled simulation with the experimental data, while the standalone *MELCOR* calculation significantly overestimates the temperature. The temperature of the inner wall of the *LIVE*-vessel is captured well by the coupled simulation. The shape of the profile matches the experimental data. The standalone *MELCOR* simulation overestimates the temperatures at intermediate height of the profile and cannot capture the high temperature close to the top of the molten material. The experiments as well as the coupled simulations showed that the behaviour of the molten material is almost independent of the applied external cooling condition. In the standalone *MELCOR* calculation a significantly higher temperature of the molten material was calculated for *LIVE*-L11 compared to *LIVE*-L10 showing a significant dependence on the cooling conditions in this code. In comparison to the standalone *MELCOR* calculation the coupled *MELCOR-PECM* system significantly improved the prediction of the behaviour of the molten material in these two experiments.

In order to show the capability of the coupled system to provide more detailed information on the progression of a severe accident in a *NPP* an entire accident was calculated. Therefore, a *MELCOR* input deck for a generic *NPP* was developed as part of this work. The postulated accident involves the occurence of a 50 cm^2 leak in the cold leg of one of the 4 primary loops. Due to the loss of coolant and the non-availability of external Severe Accident Management (*SAM*) systems, such as coolant pumps, the core melts down and relocates to the lower plenum. As soon as a constant amount of oxidic material is present in the lower plenum the calculation in the *PECM* is activated. In the coupled simulation with the *PECM* the failure of the lower head is predicted approximately 1 000 seconds after the start of the *PECM* at about 16 000 seconds of simulation time, whereas in the standalone *MELCOR* calculation the failure of the lower head is predicted after about 20 000 seconds. Comparing the calculated time since the start of the *PECM* in the coupled simulation, the simulated time until failure in standalone *MELCOR* exceeds the calculated time in the coupled simulations by a facor of 5. The significantly earlier failure has a significant impact on the possible application of *SAM* in an actual severe accident. With the coupled *PECM* also the temperature distribution and development in the entire lower plenum can be visualized.

The simulation of a severe accident in an *NPP* was improved in this work. In comparison to the original *PECM*-methodology, the developed *S-PECM* allows for the more accurate simulation of the behaviour of molten core material in the lower head, which is cooled from the top and through the lower head walls. Furthermore, the *S-PECM* is capable to resolve the stratification in the lower region of the molten material, which is not possible using the traditional *PECM*-approach. The developed coupling process between *MELCOR* and the program *DINAMO* allows for the use of new models or even entire programs such as OpenFOAM for the simulation of an *NPP*. These simulations are the first fully coupled calculations with a system code and a special code for the phenomena in the lower plenum. Due to the coupling approach the capabilities to exactly predict the physics in nuclear power plants has been improved significantly. Especially in Probabilistic Safety Analyses of Stage 2 (*PSA-2*), detailed models for specific phenomena can now be used in *MELCOR* via the coupling interface with transient boundary conditions.

Appendix

A Description of DINAMO

This chapter describes the processes in the developed coupling tool *DINAMO*. First the general execution progress of the routines to enable the coupling with MELCOR is presented. Afterwards a detailed description of the routines which form the interface for the integration of additional models is given.

A.1 Program execution

The execution of *DINAMO* is controlled by the routine called *main*. The first task of this routine is to start the *openfiles* routine to create and initialise files to track the coupling process as well as the program execution. After this step the coupling process is started by the *initialize* routine. This routine creates a Message-Passing Interface (*MPI*)-environment, which is used to establish a first communication with the other programs taking part in the coupled process and to register itself as part of the coupling in the MPIEXEC communication program. Then the parameters for the coupled calculation are determined. This is done within the *input* routine. The program *DINAMO* does not need a separate input file, because all the data needed for the upcoming coupled simulation is provided by the coupling parameters, which are defined in the input file of MPIEXEC. The coupling parameters, such as the system time or in- and output parameters of the models used in this simulation, are then sent to the routine *sortvars*. This routine analyses the received coupling parameters and assigns them to the corresponding models. Once every parameter of the desired model is defined, this specific model is activated for the further calculations, otherwise an error message is produced. A more detailed description of this routine can be found in chapter A.2. After all the models needed for the coupled calculation have been activated, the coupling process is initialised by exchanging the initial conditions using the *exchange* routine. The transient part of the coupling process is also performed by this routine. In order to start the transient calculation in the *main* routine a loop is started, which executes the *exchange* routine as long as the simulation time is below the defined end time of the calculation. The first operation in eacht coupling timestep is to transmit all data from *MELCOR*. Afterwards this data is used in the *models* routine to calculate the model parameters based on the received parameters. After the calculations in the coupled models have been performed the model output parameters are sent back to *MELCOR*. The loop in the *main* routine is stopped as soon as the termination time of the simulation is reached. In order to log the program off the coupling process the *finalize* routine is called. This routine also closes the *MPI*-environment. At the end of the program's execution the *openfiles* routine is called once again to finalise and close the files used to track the simulation. Finally, the *main* routine stops the execution and closes the program.

A.2 Assignment of variables

The routine *sortvars* is used to analyse the variables received by the *input* routine via the coupling interface and to assign these variables to the corresponding models. Therefore, in the input file of the communication program MPIEXEC the name of each parameter has to correspond to the definition of the model in *DINAMO*. The name of the model variables as well as the main calculation parameters of the models in *DINAMO* are stored in the *attributes* module. When a variable name corresponding to a model in *DINAMO* is found, the routine checks, whether all other variable names of the in- and output parameters needed by the model have been defined in the MPIEXEC input file.

If all model variables have been assigned to the corresponding coupled parameter name, the selected model is activated for the further calculation. This procedure minimises the potential for coupling errors and also reduces the calculation time of *DINAMO*, as only models needed for the coupled simulation are part of the programs execution.

A.3 Data exchange and calculations in the models

The data exchange with *MELCOR* via the communication program MPIEXEC is executed by the *exchange* routine in *DINAMO*. At the beginning of each coupling exchange the time step progress of the simulation is checked. Due to the fact that *DINAMO* does not have an own time step progress implemented, the current simulation time is identical with the MPIEXEC system time. This MPIEXEC system time also has to be simulated by *MELCOR*. When all three program show the same system time the coupling process can be executed. Therefore, first all coupling parameters from *MELCOR* have to be received by *DINAMO*. This is done in the *receive* routine in *DINAMO* (ref. figure 5.1). The received parameters are then transferred to the *models* routine, from where they are distributed according to the previous definition in the *sortvars* routine (ref capter A.2) among the models available in *DINAMO*.

Depending on the complexity of the specific model further calculations are performed directly in the *models* routine or in a model specific subroutine. An example of this process can be observed in figure A.1.

```
!----------------------------------------------------------------------!
!                                                                      !
! Execution of Model "model"                                           !
!                                                                      !
!----------------------------------------------------------------------!

! Is model used in this simulation?

          if(model) then

            if( cdebugout ) write(out,1000) modelnam
1000        format('models:        Transfering parameters to model < ',A,' >.')

! Copy model variables to an internal array

            do ivar= 1, modelvarin

              modelvar(ivar) = valuem2f(modelstartin +ivar-1)

            enddo

! Either:

! Model is run in a separate subroutine

!           call model(modelvar, gerr)

! Or:

! Model is run directly in this routine
!
! for example: --> incoming value is doubled

            modelvar(2)=modelvar(1)*2

! Copy outgoing variables to array for outgoing variables

            do ivar = 1, modelvarout

              valuef2m(modelstartout + ivar - 1) = modelvar(modelvarin + ivar)

            enddo

!----------------------------------------------------------------------!
! End of execution of model "model"                                    !
!----------------------------------------------------------------------!
```

Figure A.1: Section of the *models* subroutine

After the calculation of the model parameters based on the received input parameters from *MELCOR*, the output variables are transferred to the *exchange* routine. The transfer of the parameters from *DINAMO* via the coupling interface is performed in the *send* routine. By means of this sequential coupling (ref. chapter 2.4.1) it is possible to directly receive the reaction of the coupled model in the same coupling time step in *MELCOR*.

This approach is similar to the structure of the *MELCOR* code, in which *DINAMO* can be considered as an additional package [73]. After the coupled data from *MELCOR* has been received *DINAMO* the next coupling time step is initiated.

B Synchronisation of the coupling process

This chapter describes the realization of a synchronous coupling between *MELCOR* and *DINAMO*. The purpose is to eliminate the influence of the externally defined coupling time step in the asynchronous coupling on the calculation results in *MELCOR*.

B.1 Determination of the influence of the asynchronous coupling

In the asynchronous coupling methodology, which was used by Szabó, for example, to couple *MELCOR* with GASFLOW, the coupling time step is defined in the input file of the communication program MPIEXEC [19]. This external coupling time step has to be kept in the *MELCOR* calculation, which influences the *MELCOR* time step progress. Due to complexity of the *MELCOR* code, which includes a variety of different simplified models for different phenomena in a severe accident, even monor changes in the *MELCOR* time step progress can influence the calculation result.

Therefore, two calculations of the same problem were performed. One calculation was performed with stand-alone *MELCOR*. The other calculation was a coupled simulation between *MELCOR* and *DINAMO*. In the coupled simulation the coupled parameters were not used for the further calculation in *MELCOR* to eliminate any influence of the coupling parameters. The maximum *MELCOR* time step in both simulations is set to 0.1 s, which is identical to the coupling time step of the coupled simulation defined by MPIEXEC. The investigated scenario assumes a leakage of coolant in the hot leg of a Pressurized Water Reactor (*PWR*). During the progress of the severe accident, a part of the core material melts. The lower core support plate restrains the relocation of the molten core material to the lower plenum. Due to the thermal and physical stress on the lower core support plate, caused by the molten core material, it fails and the degraded material can relocate to the lower plenum. The two simulations are compared by the temperature profiles in the lower core support plate, which can be observed in figure B.1.

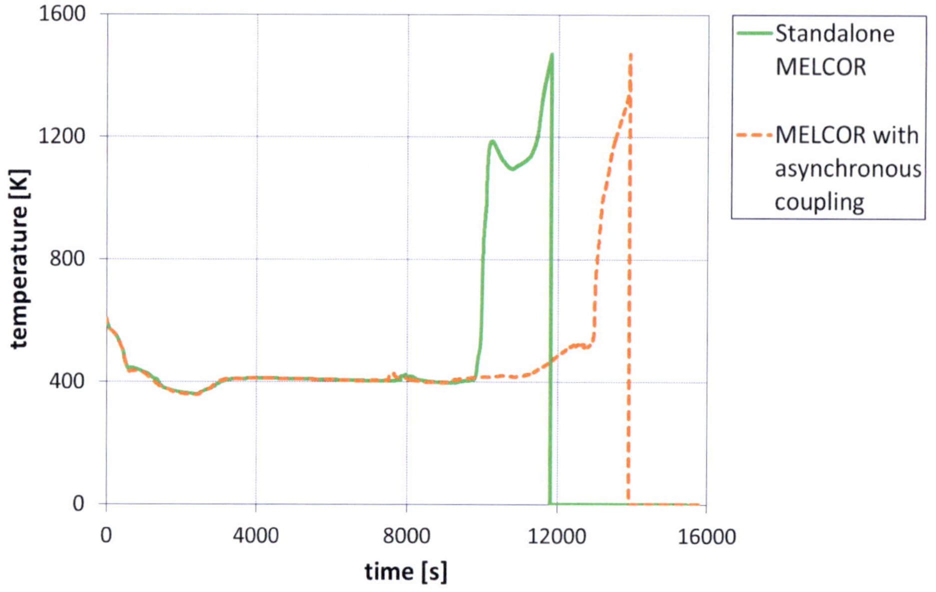

Figure B.1: Comparison of the temperature development in the lower core support plate to determine the influence of the coupling process

The temperature development in the lower core support plate in the stand-alone *MELCOR* calculation shows at the beginning of the simulation a high congruency with the temperature profile in the asynchronously coupled simulation. At about 10 000 seconds a significant difference in the temperature development is observed. This induces an approximately 2000 seconds earlier failure of the lower core support plate in the stand-alone simulation. These significant deviances are caused by the different *MELCOR* time step progress in the asynchronous coupling approach. A comparison of the different time step sizes in the *MELCOR*-simulation is shown in figure B.2.

The time step in the stand-alone calculation differs significantly from the time step in the asynchronous coupling. Due to the external definition of the coupling time step the *MELCOR* time step has to be reduced to reach the required system time exactly. In this work a synchronous coupling methodology is developed in order to eliminate the influence of the coupling process on the *MELCOR* calculation.

B.2 Development of a synchronous coupling

In order to realize a synchronous coupling between *MELCOR* and *DINAMO*, using MPIEXEC as a communication program, the source codes of all three programs have to be modified. As described in chapter 2.4 ,in a synchronous coupling the time step progress of the coupled

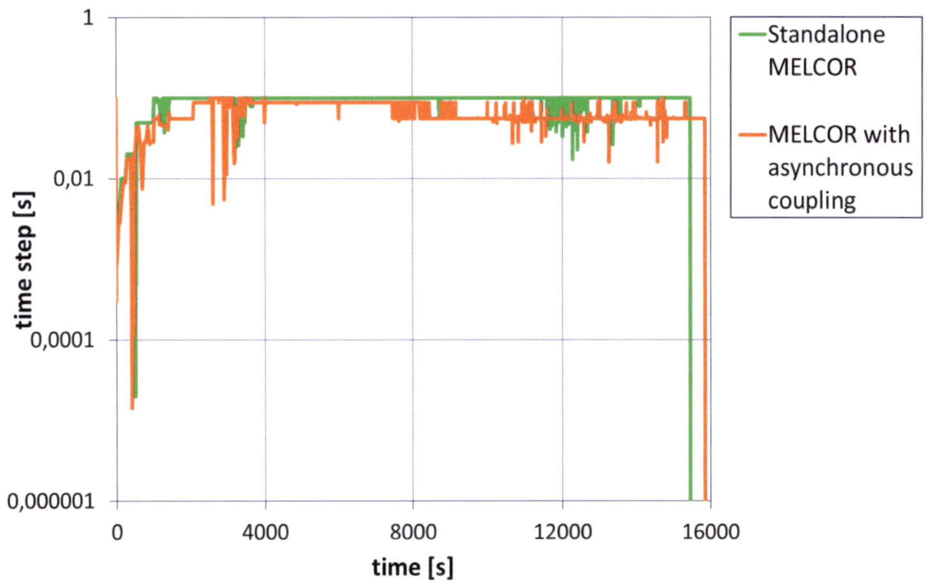

Figure B.2: Comparison of the *MELCOR* time step sizes to determine the influence of the coupling process

programs is identical. In this special case the coupling time step, defined by MPIEXEC, and the system time step in *MELCOR* are synchronized to eliminate the influence of the coupling process on the *MELCOR* simulation. The intention of the synchronous coupling process is that the coupling is no longer executed at the predefined time steps from the MPIEXEC input file but now corresponds to the system time step in *MELCOR*. In *DINAMO* the changing of the coupling time step has no influence, because the time step in *DINAMO* in the asynchronous coupling was already based on the external coupling time step. In figure B.3 a schematic of the synchronous time step progress can be found.

At the beginning of each system time step in *MELCOR* the current system time in all three programs (*MELCOR*, MPIEXEC and *DINAMO*) is sent to MPIEXEC (t_M = System time in *MELCOR*; t_{MPI} = System time in MPIEXEC; t_D = System time in *DINAMO*). The minimum system time t_{min} is determined and set as the new system time for all three programs. After the synchronization of the system times the data exchange is performed as described in chapter A.3. After the coupling exchange, *MELCOR* simulates the next system time step based on its time step progress routine (dt_M = Time step in *MELCOR*). MPIEXEC and *DINAMO* on the other hand also advance according to their time step progress, which is based on the coupling time step defined in the MPIEXEC input file (dt_{MPI} = Time step in MPIEXEC; dt_D = Time step in *DINAMO*). This coupling time step in MPIEXEC has to be defined in such a way that it always exceeds the system time step of *MELCOR* dt_M. This methodology ensures that the new system time in *MELCOR* ($t_M{}^*$) at the end of each time step of the coupled system is always smaller than the new system times in MPIEXEC ($t_{MPI}{}^*$) and *DINAMO* ($t_D{}^*$). Therefore, the system time of the coupled system directly corresponds to the system time in *MELCOR*. This approach eliminates the influence of the asynchronous coupling. Furthermore, a closer connection between the coupled programs is realized, because now at every *MELCOR* system time step a coupling exchange is performed.

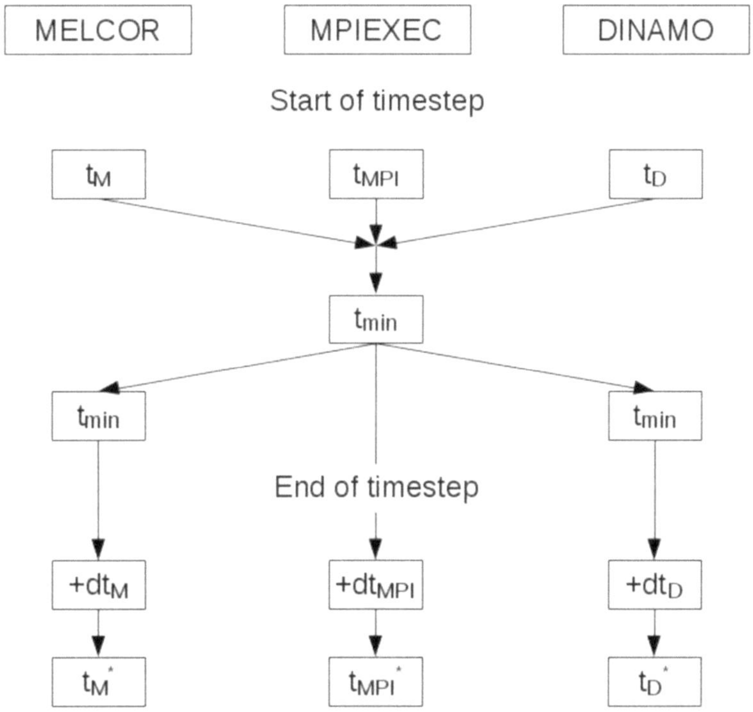

Figure B.3: Schematic of the synchronous time step progress

The temperature development in the lower core support plate for the hypothetical severe accident introduced in chapter B for the synchronous coupling and the stand-alone *MELCOR* calculation can be found in figure B.4.

The temperatures in the lower core support plate for the stand-alone and the synchronously coupled simulation are now identical. This is acheived by the different time step progress in the synchronous coupling in contrast to the asynchronous coupling. As shown in figure B.5 the time steps in the synchronous coupling and the stand-alone calculation are also exactly the same. With the synchronous coupling methodology the effect of the coupling process on the *MELCOR* calculation is eliminated. Furthermore, due to the synchronous coupling the connection between *MELCOR* and the coupled models in *DINAMO* is improved.

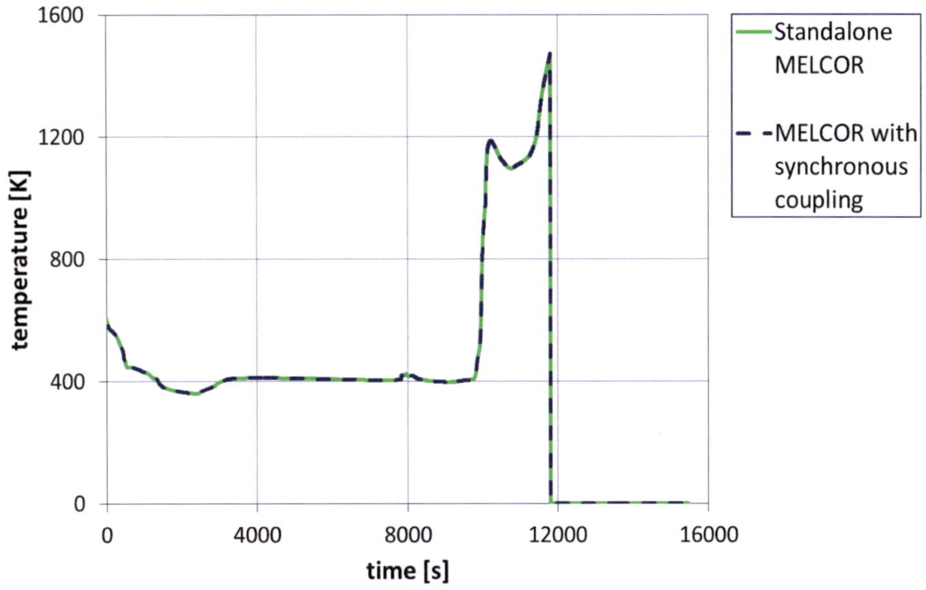

Figure B.4: Comparison of the temperature in the lower core support plate using the synchronized coupling process

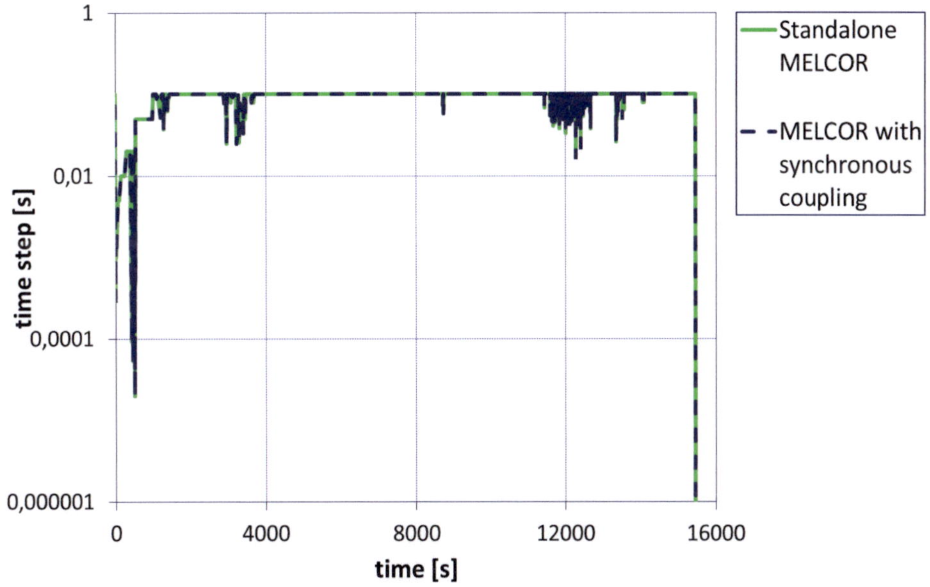

Figure B.5: Comparison of the *MELCOR* time step progress between the stand-alone *MELCOR* calculation and the synchronously coupled simulation

Bibliography

[1] D. Goluskin, *Internally heated convection and Rayleigh-Bénard convection.* Springer International Publishing, 2016.

[2] http://www.weltraumport.de.

[3] H. Oertel and J. Delfs, *Strömungsmechanische Instabilitäten : mit 82 Abbildungen.* Springer, 1996.

[4] R. Müller, "Observations of the solar granulation at the pic du midi observatory and with themis," in *Solar and Stellar Granulation*, ser. NATO ASI Series, R. Rutten and G. Severino, Eds. Springer Netherlands, 1989, vol. 263, pp. 9–12.

[5] http://www.mantleplumes.org.

[6] H. Lugt, *Wirbelströmung in Natur und Technik.* Braun, 1979.

[7] T. G. Theofanous, C. Liu, S. Additon, S. Angelini, O. Kymäläinen, and T. Salmassi, "In-Vessel Coolability and Retention of a Core Melt," University of California, Santa Barbara, Tech. Rep., 1996.

[8] E. L. Tolman, P. Kuan, and J. M. Broughton, "TMI-2 Accident Scenario Update," *Nuclear Engineering and Design*, vol. 108, pp. 45 – 54, 1988.

[9] B. R. Sehgal *et al.*, *Nuclear Safety in Light Water Reactors - Severe Accident Phenominology*, B. R. Sehgal, Ed. Academic Press, 2012.

[10] P. Tusheva, "Modelling and Analysis of Severe Accidents for VVER-1000 Reactors," Ph.D. dissertation, Technische Universität Dresden, 2011.

[11] P. Laufs, *Reaktorsicherheit für Leistungskernkraftwerke - Die Entwicklung im politischen und technischen Umfeld der Bundesrepublik Deutschland.* Springer VIEWEG, 2013.

[12] G. Kessler *et al.*, *Sicherheit von Leichtwasserreaktoren - Risiken der Nukleartechnologie.* Springer Verlag,, 2012.

[13] L. Zhang, Y. Zhou, Y. Zhang, W. Tian, S. Qiu, and G. Su, "Natural convection heat transfer in corium pools: A review work of experimental studies," *Progress in Nuclear*

Energy, vol. 79, pp. 167 – 181, 2015.

[14] A. Palagin, A. Miassoedov, X. Gaus-Liu, M. Buck, C. T. Than, P. Kudinov, L. Carenini, C. Koellein, W. Luther, and V. Chudanov, "Analysis and Interpretation of the LIVE-L6 Experiment," *Proceedings of the 5th European Review Meeting on Severe Accident Research (ERMSAR-2012), Cologne, Germany, March 21 - 23*, 2012.

[15] C. T. Tran, "The Effective Convectivity Model for Simulation and Analysis of Melt Pool Heat Transfer in a Light Water Reactor Pressure Vessel Lower Head," Ph.D. dissertation, Royal Institute of Technology (KTH), 2009.

[16] B. Fluhrer, H. Alsmeyer, T. Cron, G. Messemer, A. Miassoedov, and T. Wenz, "The Experimental Programme LIVE to Investigate In-Vessel Core Melt Behaviour in the Late Phase," *Proceedings of the Jahrestagung Kerntechnik*, 2005.

[17] R. Gauntt, "7th MELCOR Users' Workshop," *Albuquerque NM, September 26 - 30*, 2005.

[18] P. Chatelard and N. Reinke, "Overview of the integral code ASTEC V2.0," French Institute for Radiological Protection and Nuclear Safety (IRSN) [Rapport Technique DPAM/SEMCA-2009-149] and Gesellschaft für Anlagen und Reaktorsicherheit (GRS) [Technische Notiz GRS ASTEC 09/02], Tech. Rep., 2009.

[19] T. Szabó, "Verbesserte Berechnung der Wasserstoffverteilung im Sicherheitsbehälter bei der Analyse schwerer Störfälle in Kernkraftwerken durch Kopplung von MELCOR und GASFLOW," Ph.D. dissertation, Karlsruhe Institute of Technology (KIT), 2013.

[20] M. Viellieber, P. Dietrich, and A. Class, "Investigation of a Wire Wrapped Fuel Assembly with the Anisotropic Coarse-Grid-CFD (AP-CGCFD)," *International Journal for Nuclear Power*, vol. 10, pp. 573 – 575, October 2013.

[21] H.-G. Willschütz, E. Altstadt, B. R. Sehgal, and F.-P. Weiss, "Recursively coupled thermal and mechanical FEM-analysis of lower plenum creep failure experiments," *Annals of Nuclear Energy*, vol. 33, pp. 126 – 148, 2006.

[22] R. Heuss, "Thermisches Versagen von Reaktordruckbehältern bei extremen Störfällen in Druckwasserreaktoren - Analyse und Verbesserungsvorschläge," Ph.D. dissertation, Rheinisch-Westfälische Technische Hochschule (RWTH) Aachen, 2005.

[23] International Nuclear Safety Advisory Group, "Defence in depth in Nuclear Safety, INSAG-10," International Atomic Energy Agency, Vienna, Tech. Rep., 1996.

[24] "Safety Reports Series No. 23 - Accident Analysis for Nuclear Power Plants," International Atomic Energy Agency, Vienna, Tech. Rep., 2002.

[25] http://www.theenergycollective.com.

[26] "Severe Accident Management Programmes for Nuclear Power Plants, Safety Guide No. NS-G-2.15," International Atomic Energy Agency, Vienna, Tech. Rep., 2009.

[27] "IAEA Safety Standards - Safety of Nuclear Power Plants: Design - Specific Safety Requirements No. SSR-2/1," International Atomic Energy Ageny, Vienna, Tech. Rep., 2012.

[28] "IAEA Safety Standards - Safety of Nuclear Power Plants: Design - Requirements No. NS-R-1," International Atomic Energy Ageny, Vienna, Tech. Rep., 2000.

[29] "Safety Reports Series No. 32 - Implementation of Accident Management Programmes in Nuclear Power Plants," International Atomic Energy Agency, Tech. Rep., 2004.

[30] A. Bentaïb *et al.*, "Research and development with regard to severe accidents in pressurised water reactors: Summary and outlook," French Institute for Radiological Protection and Nuclear Safety (IRSN) and French Atomic Energy Commission (CEA), Tech. Rep. Rapport IRSN-2007/83 and Rapport CEA-2007/351, 2007.

[31] B. R. Sehgal, "Stabilization and termination of severe accidents in LWRs," *Nuclear Engineering and Design*, vol. 236, pp. 1941 – 1952, 2006.

[32] B. R. Sehgal, P. Piluso *et al.*, "SARNET Lecture Notes on Nuclear Reactor Severe Accident Phenominology," French Atomic Energy Agency (CEA), Tech. Rep., 2008.

[33] J. L. Rempe, K. Y. Suh, F. B. Cheung, and S. Kim, "In-Vessel Retention of Molten Corium: Lessons Learned and Outstanding Issues," *Nuclear Technology*, vol. 161, no. 3, pp. 210 – 267, March 2008.

[34] O. Kymäläinen, H. Tuomisto, and T. G. Theofanous, "In-vessel retention of corium at the Loviisa plant," *Nuclear Engineering and Design*, pp. 109 – 130, 1997.

[35] "Considerations concerning the strategy of corium retention in the reactor vessel," French Institute for Radiological Protection and Nuclear Safety (IRSN), Tech. Rep., 2015.

[36] J.-M. Seiler, A. Latrobe, B. Sehgal, H. Alsmeyer, O. Kymäläinen, B. Turland, J.-L. Grange, M. Fischer, G. Azarian, M. Bürger, C. Cirauqui, and A. Zurita, "Analysis of corium recovery concepts by the EUROCORE group," *Nuclear Engineering and Design*, vol. 221, no. 1Ű3, pp. 119 – 136, 2003, mid-Term Symposium on Shared-Cost and Concerted Actions.

[37] T. G. Theofanous and S. Angelini, "Natural convection for in-vessel retention at prototypic Rayleigh numbers," *Nuclear Engineering and Design*, vol. 200, pp. 1 – 9,

2000.

[38] T. G. Theofanous, M. Maguire, S. Angelini, and T. Salmassi, “The first results from the ACOPO experiment,” *Nuclear Engineering and Design*, vol. 169, pp. 49 – 57, 1997.

[39] J. M. Bonnet and J. M. Seiler, “Thermal Hydraulic Phenomena in Corium Pools: The BALI Experiment,” *Proceedings of the 7th International Conference on Nuclear Engineering (ICONE-7), Tokyo, Japan, April 19 - 23*, 1999.

[40] B. R. Sehgal *et al.*, “Assesment of reactor vessel integrity,” *Nuclear Engineering and Design*, vol. 235, pp. 213 – 232, 2005.

[41] O. Kymäläinen., H. Tuomisto, O. Hongisto, and T. G. Theofanous, “Heat flux distribution from a volumetrically heated pool with high Rayleigh number,” *Nuclear Engineering and Design*, vol. 149, pp. 401 – 408, 1994.

[42] M. Helle, O. Kymäläinen, and H. Tuomisto, “Experimental COPO II Data on Natural Convection in Homogenious and Stratified Pools.” *Proceedings of the 9th International Topical Meeting on Nuclear Reactor Thermal Hydraulics (NURETH-9), San Francisco, California, October 3 - 8*, 1999.

[43] B. R. Sehgal, A. Giri, U. Chikkanagoudar, and A. Karbojian, “Experiments on in-vessel melt coolability in the EC-FOREVER Program,” *Nuclear Engineering and Design*, vol. 236, pp. 2499 – 2210, 2006.

[44] R. Azizian, T. McKrell, K. Atkhen, and J. Buongiorno, “Effects of porous superhydrophilic Surfaces on flow boiling Critical Heat Flux in IVR Accident Scenarios,” *Proceedings of the 16th International Topical Meeting on Nuclear Reactor Thermalhydraulics, Chicago, USA, August 30 - September 4*, 2015.

[45] V. G. Asmolov, S. S. Abalin, Y. G. Degal’tsev, O. Y. Shakh, E. K. D’yakov, and V. F. Strizhov, “Behavior of a Core Melt Pool on the Floor of the Reactor Vessel (RASPLAV Project),” *Atomic Energy*, vol. 84, no. 4, pp. 241 – 255, 1998.

[46] V. Asmolov, S. Abalin, A. Surenkov, I. Gnidoi, and V. Strizhov, “Results of Salt Experiments Performed during Phase I of RASPLAV Project,” Institute of Nuclear Safety (IBRAE), Russian Academy of Science, Tech. Rep., 1998.

[47] B. R. Sehgal, V. A. Bui, T. N. Dinh, J. A. Green, and G. Kolb, “SIMECO Experiments on In-Vessel Melt Pool Formation and Heat Transfer with and without a Metallic Layer,” *Proceedings of the OECD Workshop on In-Vessel Core Debris Retention and Coolability, Garching, Germany*, 1998.

[48] A. V. Stepanyan, A. K. Nayak, and B. R. Sehgal, “Experimental Investigation of natural

Convection in a three Layer stratified Pool with internal Heat Generation," *Proceedings of the 11th International Topical Meeting on Nuclear Reactor Thermal Hydraulics (NURETH-11), Popes' Palace Conference Center, Avignon, France, October 2 - 6*, 2005.

[49] B. R. Sehgal and H. S. Park, "Pre-Project on Development and Validation of Melt Behavior in Severe Accidents," Royal Institute of Technology (KTH), Tech. Rep., 2004.

[50] S. A. Theerthan, G. Kolb, and B. R. Sehgal, "The Role of Phase Separation on Heat Transfer in Internally Heated Liquid Layers," *Proceedings of the 8th International Converence on Nuclear Engineering (ICONE-8), Baltimore, USA, April 2 - 6*, 2000.

[51] S. Rougé, "SULTAN test facility for large-scale vessel coolability in natural convection at low pressure," *Nuclear Engineering and Design*, vol. 169, pp. 185 – 195, 1997.

[52] A. Miassoedov, X. Gaus-Liu, T. Cron, and B. Fluhrer, "LIVE Experiments on Melt Pool Heat Transfer in the Reactor Pressure Vessel Lower Head," *Proceedings of the 10th International Conference on Heat Transfer, Fluid Mechanics and Thermodynamics HEFAT2014), Orlando, Florida, 14 - 26 July*, 2014.

[53] F. Kretzschmar and B. Fluhrer, "Behavior of the Melt Pool in the Lower Plenum of the Reactor Pressure Vessel - Review of Experimental Programs and Background of the LIVE Program," Forschungszentrum Karlsruhe, FZKA 7382, Tech. Rep., 2008.

[54] A. Miassoedov, T. Cron, J. Foit, S. Schmidt-Stiefel, T. Wenz, I. Ivanov, and D. Popov, "Results of the LIVE-L1 Experiment on Melt Behaviour in RPV Lower Head performed within the LACOMERA Project at the Forschungszentrum Karlsruhe," *Proceedings of the 15th International Conference on Nuclear Engineering (ICONE15), Nagoya, Japan, April 22 - 26*, 2007.

[55] R. W. Berg and D. H. Kerridge, "The NaNO3-KNO3 system: the position of solidus and sub-solidus," *Dalton Transactions (Print Edition)*, no. 15, pp. 2224–2229, 2004.

[56] X. Gaus-Liu, A. Miassoedov, T. Cron, J. Foit, B. Fluhrer, S. Schmidt-Stiefel, and T. Wenz, "Test and simulation results of LIVE-L4 + LIVE-L5L," Karlsruhe Institute of Technology (KIT), KIT Scientific Reports 7593, Tech. Rep., 2011.

[57] A. Miassoedov, T. Cron, J. Foit, B. Fluhrer, S. Schmidt-Stiefel, and T. Wenz, "Results of the LIVE-I1 commissioning test on molten pool behaviour in the lower head of the RPV," *Proceedings of the Jahrestagung Kerntechnik, Karlsruhe, May 22 - 24*, 2007.

[58] B. Fluhrer, A. Miassoedov, T. C. J. Foit, X. Gaus-Liu, S. Schmidt-Stiefel, T. Wenz, I. Ivanov, and D. Popov, "The LIVE-L1 and LIVE-L3 Experiments on Melt Behaviour in RPV Lower Head," Forschungszentrum Karlsruhe, FZKA 7419, Tech. Rep., 2008.

[59] A. Miassoedov, T. Cron, and J. Foit, "Results of the LIVE-L2 experiment on melt behaviour in RPV lower head performed within the LACOMERA project at the Forschungszentrum Karlsruhe," *Proceedings of the 19th International Conference on Structural Mechanics in Reactor Technology (SMiRT-19), Toronto, Canada, August 12 - 17*, 2007.

[60] X. Gaus-Liu, A. Miassoedov, J. Foit, T. Cron, F. Kretzschmar, T. Wenz, and S. Schmidt-Stiefel, "LIVE-L4 and LIVE-L5L Experiments on Melt Pool and Crust Behaviour in the RPV Lower Head," *Proceedings of the 14th International Topical Meeting in Nuclear Reactor Thermal Hydraulics (NURETH-14), Toronto, Canada, September 25 - 29*, 2011.

[61] X. Gaus-Liu, A. Miassoedov, T. Cron, and B. Fluhrer, "LIVE-3D and LIVE-2D Experimental Results on Melt Pool Heat Transfer With Top Cooling Conditions," *Proceedings of the 6th European Review Meeting on Severe Accident Research (ERMSAR-2013), Avignon, France, October 2 - 4*, 2013.

[62] X. Gaus-Liu, A. Miassoedov, T. Cron, S. Schmidt-Stiefel, and T. Wenz, "The Experimental Results of LIVE-L8B - Debris MelMelt Process in a Simulated PWR Lower Head," *Proceedings of the 5th European Review Meeting on Severe Accident Research (ERMSAR-2012), Cologne, Germany, March 21 - 23*, 2012.

[63] A. Miassoedov, T. Cron, X. Gaus-Liu, A. Palagin, S. Schmidt-Stiefel, and T. Wenz, "LIVE Experiments on Melt Behavior in the Reactor Pressure Vessel Lower Head," *Heat Transfer Engineering*, vol. 34, no. 14, pp. 1226 – 1236, 2013.

[64] P. Dietrich, X. Gaus-Liu, F. Kretzschmar, A. Miassoedov, and A. G. Class, "LIVE-Experiments on the In-Vessel Melt Pool Behaviour with Different External Cooling Conditions and the Simulation Results with a coupled MELCOR-PECM Approach," *Proceedings of the 16th International Topical Meeting on Nuclear Reactor Thermalhydraulics, Chicago, USA, August 30 - September 4*, 2015.

[65] F. Mayinger, M. Jahn, H. H. Reineke, and U. Steinberner, "Untersuchung thermohydraulischer Vorgänge sowie Wärmeaustausch in der Kernschmelze," Technische Universität Hannover, Tech. Rep., 1975.

[66] F. A. Kulacki and R. J. Goldstein, "Thermal convection in a horizontal fluid layer with uniform volumetric energy sources," *Journal of Fluid Mechanics*, vol. 55, pp. 271–287, 9 1972.

[67] P. Bachmann, C. Müller, J. Sievers, and T. Schimpfke, "Development and Verification of Model for the Simulation of Debris/RPV Wall Interaction," Gesellschaft für Anlagen

und Reaktorsicherheit (GRS), Final Report for Reactor Safety Research Project No.: RS 1002, Tech. Rep., 2001.

[68] U. Steinberner and H. H. Reineke, "Turbulent buoyancy convection heat transfer with internal heat sources," *Energy: The international Journal*, vol. 2, 1978.

[69] F. J. Asfia, B. Frantz, and V. Dhir, "Experimental investigation of natural convection heat transfer in volumetrically heated spherical segments," *Journal of Heat Transfer*, vol. 118, no. 1, pp. 31 – 37, 1996.

[70] J. D. Gabor, P. G. Ellison, and J. C. Cassulo, "Heat Transfer from Internally Heated Hemispherical Pools," *Proceedings of the 19th National Heat Transfer Conference, Orlando, Florida, USA, July*, 1980.

[71] D. Dropkin and E. Somerscales, "Heat transfer by natural convection in liquids confined by two parallel plates which are inclined at various angles with respect to the horizontal," *Journal of Heat Transfer*, vol. 87, no. 1, pp. 77 – 82, 1965.

[72] S. W. Churchill and H. H. Chu, "Correlating equations for laminar and turbulent free convection from a vertical plate," *International Journal of Heat and Mass Transfer*, vol. 18, no. 11, pp. 1323 – 1329, 1975.

[73] R. O. Gauntt, J. E. Cash, R. K. Cole, C. M. Erickson, L. L. Humphries, S. B. Rodriguez, and M. F. Young, "MELCOR Computer Code Manuals - Vol. 2 Reference Manuals - Version 1.8.6 September," Sandia National Laboratories (SNL), Tech. Rep., 2005.

[74] T. C. Chawla and S. Chan, "Heat Transfer from Vertical/Inclined Boundaries of Heat-Generating Boiling Pools," *Journal of Heat Transfer*, vol. 104, pp. 465 – 473, 1982.

[75] E. R. G. Eckert and T. W. Jackson, "Report 1015 - Analysis of Turbulent Free-Convection Boundary Layer on Flat Plate," National Advisory Dommitee for Aeronautics, Tech. Rep., 1950.

[76] W. Kurz and D. Fisher, "Fundamentals of solidification, 1986," *Trans Tech Publications, Switzerland.*

[77] W. Boettinger, S. Coriell, A. Greer, A. Karma, W. Kurz, M. Rappaz, and R. Trivedi, "Solidification microstructures: recent developments, future directions," *Acta Materialia*, vol. 48, no. 1, pp. 43 – 70, 2000.

[78] C. T. Tran and T. N. Dinh, "The effective convectivity model for simulation of melt pool heat transfer in a light water reactor pressure vessel lower head. Part I: Physical processes, modmodel and model implementation," *Progress in Nuclear Energy*, vol. 51, pp. 849 – 859, 2009.

[79] F. B. Cheung, S. W. Shiah, D. H. Cho, and M. J. Tan, "Model of heat transfer in a horizontal heat-generating layer by an effective diffusivity approach," *Proceedings of the National Heat Transfer Conference, San Diego, USA, August 9 - 12*, 1992.

[80] SCDAP/RELAP5-3D Code Development Team, "SCDAP/RALAP5-5D Code Manual - Volume 2: Modeling of Reactor Core and Vessel Behaviour during Severe Accidents," Idaho National Engineering and Enviromental Laboratory, Tech. Rep., 2003.

[81] F. Kulacki and M. Nagle, "Natural convection in a horizontal fluid layer with volumetric energy sources," *Journal of Heat Transfer*, vol. 97, no. 2, pp. 204 – 211, 1975.

[82] V. A. Bui and T. N. Dinh, "Modelling of Heat Transfer in Heat-Generating Liquid Pools by an Effective Diffusivity Convectivity Approach," *Proceedings of the 2nd European Thermal-Sciences and 14th UIT National Heat Transfer Conference*, 1996.

[83] B. R. Sehgal, T. D. V. A. Bui, and R. R. Nourgaliev, "Heat Transfer Processes in Reactor Vessel Lower Plenum during Late Phase of In-Vessel Core Melt Progression," *Advances in Nuclear Science and Technology*, vol. 26, pp. 103 – 135, 1999.

[84] F. Kulacki and A. Emara, "Steady and transient thermal convection in a fluid layer with uniform volumetric energy sources," *Journal of Fluid Mechanics*, vol. 83, no. 02, pp. 375 – 395, 1977.

[85] C. T. Tran and T. N. Dinh, "Simulation of Core Melt Pool Formation in a Reactor Pressure Vessel Lower Head using an Effective Convectivity Model," *Nuclear Engineering and Technology*, vol. 41, no. 7, pp. 929 – 944, 2009.

[86] L. Bernaz, J. M. Bonnet, B. Spindler, and C. Villermaux, "Thermalhydraulic Phenomena in Corium Pools - Numerical Simulation with TOLBIAC and Experimental Validation with Bali," *Workshop on in-vessel core debris retention and coolability, Garching, Germany, March 3 - 6*, 1998.

[87] C. T. Tran and T. N. Dinh, "The effective convectivity model for simulation of melt pool heat transfer in a light water reactor pressure vessel lower head. Part II: Model assessment and application," *Progress in Nuclear Energy*, vol. 51, pp. 860 – 871, 2009.

[88] C. T. Tran and P. Kudinov, "The Effective Convectivity Model for Simulation of Molten Metal Layer Heat Transfer in a Boiling Water Reactor Lower Head," *Science and Technology of Nuclear Installations*, 2013.

[89] "Safety Reports Series No. 56 - Approaches and Tools for Severe Accident Analysis for Nuclear Power Plants," International Atomic Energy Agency, Tech. Rep., 2008.

[90] W. Klein-HeSSling, N. Reinke, H. W. Chan, and D. Krönung, "Analyse kernzerstörungs

und quelltermrelevanter Prozesse mit ASTEC und ATHLET-CD/COCOSYS," Gesellschaft für Anlagen und Reaktorsicherheit (GRS), GRS-351, Tech. Rep., 2014.

[91] "Mitigation of Hydrogen Hazards in Severe Accidents in Nulcear Power Plants," International Atomic Energy Agency (IAEA), IAEA-TECDOC-1661, Tech. Rep., 2011.

[92] D. Baraldi, M. Heitsch, and H. Wilkening, "CFD simulations of hydrogen combustion in a simplified EPR containment with CFX and REACFLOW," *Nuclear Engineering and Design*, vol. 237, no. 15, pp. 1668 – 1678, 2007.

[93] R. O. Gauntt, J. E. Cash, R. K. Cole, C. M. Erickson, L. L. Humphries, S. B. Rodriguez, and M. F. Young, "MELCOR Computer Code Manuals - Vol. 1 Primer and Users' Guide - Version 1.8.6 September," Sandia National Laboratories (SNL), Tech. Rep., 2005.

[94] K. Voelsing, "Modular Accident Analysis Program 5 (MAAP5) Application Guidance: Desktop Reference for Using MAAP5 Software - PPhase 2 Report," EPRI, Palo Alto, CA, 3002005285, Tech. Rep., 2015.

[95] L. L. Humphries, R. Gauntt, and H. Esmaili, "MELCOR Code Development Status, Code Assessment, and QA," *1st Meeting of the "European MELCOR User Group" (EMUG), Villigen, Switzerland, December 15 - 16*, 2008.

[96] https://melcor.sandia.gov.

[97] U. Brockmeier, "Severe core damage model in the code ATHLET-CD, ICARE2, MELCOR and SCDAP/RELAP," European Commission, EUR 16937 DE/EN, Tech. Rep., 1996.

[98] W. Klein-HeSSling, S. Arndt, G. Weber, H. Wolff, S. Schwarz, and C. Spengler, "COCOSYS - Short description," Gesellschaft für Anlagen und Reaktorsicherheit (GRS), Tech. Rep., 2008.

[99] R. K. Cole, "Coupling of MELCOR to Other CCode under an Executive Program using PVM Message Exchange," *Prodeedings of the 2002 RELAP5 Users Seminar, Park City, Utah, September 4 - 6*, 2002.

[100] P. Dietrich, F. Kretzschmar, and A. Miassoedov, "Verbesserungen von Lower Head Modellen in MELCOR - Auswahl und Integration neuer Modelle," Karlsruhe Institute of Technology (KIT), Technischer Fachbericht zum Meilenstein M22 des Verbundvorhabens WASA-BOSS, Tech. Rep., 2014.

[101] K. V. Moore and W. H. Rettig, "RELAP 2- A digital program for reactor blowdown and power excursion analysis (Digital program for reactor blowdown and power excursion

analysis)," 1968.

[102] D. Pialla, D. Tenchine, S. Li, P. Gauthe, A. Vasile, R. Baviere, N. Tauveron, F. Perdu, L. Maas, F. Cocheme, K. Huber, and X. Cheng, "Overview of the system alone and system/cfd coupled calculations of the {PHENIX} natural circulation test within the {THINS} project," *Nuclear Engineering and Design*, vol. 290, pp. 78 – 86, 2015, thermal-Hydraulics of Innovative Nuclear Systems.

[103] T. Schenkel, S. Krittian, K. Spiegel, S. Höttges, M. Perschall, and H. Oertel, *World Congress on Medical Physics and Biomedical Engineering, September 7 - 12, 2009, Munich, Germany: Vol. 25/4 Image Processing, Biosignal Processing, Modelling and Simulation, Biomechanics.* Berlin, Heidelberg: Springer Berlin Heidelberg, 2010, ch. The Karlsruhe Heart Model KaHMo: A modular framework for numerical simulation of cardiac hemodynamics, pp. 615–618.

[104] S. Guo, J. Xu, J. Mo, R. Gu, and L. Pang, "FluidŰstructure interaction study of the splitter plate in a {TBCC} exhaust system during mode transition phase," *Acta Astronautica*, vol. 112, pp. 126 – 139, 2015.

[105] Y. Yamamoto and M. Yoshioka, "Cfd and fem coupling analysis of orex aerothermodynamic flight data," *AIAA Paper*, vol. 95, p. 2087, 1995.

[106] M. Christian, L. P. Françoise, and A. Ludovic, "A least-squares coupling method between a finite element code and a discrete element code," *International Journal for Numerical Methods in Engineering*, vol. 101, no. 10, pp. 731–743, 2015.

[107] S. Akbas, V. Martinez-Quiroga, F. Aydogan, A. M. Ougouag, and C. Allison, "Survey of coupling schemes in traditional coupled neutronics and thermal-hydraulics codes," in *ASME 2015 International Mechanical Engineering Congress and Exposition.* American Society of Mechanical Engineers, 2015, pp. V06BT07A027–V06BT07A027.

[108] J. Kim and S.-W. Hong, "Analysis of hydrogen flame acceleration in {APR1400} containment by coupling hydrogen distribution and combustion analysis codes," *Progress in Nuclear Energy*, vol. 78, pp. 101 – 109, 2015.

[109] C. Chauliac, J.-M. Aragonés, D. Bestion, D. G. Cacuci, N. Crouzet, F.-P. Weiss, and M. A. Zimmermann, "{NURESIM} Ű a european simulation platform for nuclear reactor safety: Multi-scale and multi-physics calculations, sensitivity and uncertainty analysis," *Nuclear Engineering and Design*, vol. 241, no. 9, pp. 3416 – 3426, 2011, seventh European Commission conference on Euratom research and training in reactor systems (Fission Safety 2009).

[110] http://www.nuresafe.eu.

[111] M. N. Avramova and K. N. Ivanov, “Verification, validation and uncertainty quantification in multi-physics modeling for nuclear reactor design and safety analysis,” *Progress in Nuclear Energy*, vol. 52, no. 7, pp. 601 – 614, 2010.

[112] D. Grigic and T. Kozlowski, “Basis for Coupling 3D Neutro-Kinetics/Thermal-Hydraulics Codes,” *Seminar and Training on Scaling, Uncertainty and 3D Coupled Calculations in Nuclear Technology, Royal Institute of Technology (KTH), Stockholm, Sweden*, 2009.

[113] K. Ivanov and M. Avramova, “Challenges in coupled thermal–hydraulics and neutronics simulations for LWR safety analysis,” *Annals of Nuclear Energy*, vol. 34, no. 6, pp. 501 – 513, 2007.

[114] K. S. Ud-Din, M. Peng, and M. Zubair, “Neutronics and thermal hydraulic coupling methods for the nuclear reactor core,” *Proceedings of the Power and Energy Engineering Conference (APPEEC), Asia-Pacific*, pp. 1 – 4, 2011.

[115] “CRISSUE-S-WP2 - Neutronics - Thermal-hydraulics Coupling in LWR Technology: State-of-the-art Report (REAC-SOAR),” Nuclear Engergy Agency - Organisation for Economic Co-operation and Development, Tech. Rep., 2004.

[116] W. L. Weaver, “The Application Programming Interface For The PVMEXEC Program And Associated Code Coupling System,” Idaho National Laboratories (INL), Tech. Rep., 2005.

[117] W. Weaver, E. Tomlinson, and D. Aumiller, “An Executive Program for use with RELAP5-3D©,” *Proceedings of the 2001 RELAP5 Users Seminar, Sun Valley, Idaho*, 2001.

[118] A. Geist, A. Beguelin, J. Dongarra, W. Jiang, R. Manchek, and V. Sunderam, *PVM: Parallel Virtual Machine - A Users' Guide and Tutorial for Networked Parallel Computing*. The MIT Press - ISBN 0-262-57108-0, 1994.

[119] W. Gropp, E. Lusk, and A. Skjellum, *MPI - Eine Einführung - Portable parallele Programierung mit dem Message Passing Interface*. Oldenbourg Verlag München Wien, 2007.

[120] The RELAP5-3D© Code Development Team, “RELAP5-3D© Code Manual Volume I - V Revision 2.4,” Idaho National Laboratory (INL), Idaho Falls, Idaho, Tech. Rep., 2005.

[121] W. L. Weaver, “Programmers Manual for the PVM Coupling Interface in the RELAP5-3D© Code,” Idaho National Laboratory (INL), Idaho Falls, Idaho, INL/EXT-05-00203, Tech. Rep., 2005.

[122] ——, "Programmers Manual for the PVMEXEC Program," Idaho National Laboratory (INL), Idaho Falls, Idaho, INL/EXT-05-00159, Tech. Rep., 2005.

[123] L. Oriani, L. E. Conway, D. Grgic, T. Bajs, M. E. Ricotti, and A. Barroso, "Overview of Computational Challenges in the Development of Evaluation Model for Safety Analyses," *Proceedings of the Nuclear Mathematical and Computational Sciences: A Century in Review, A Century Anew, Gatlinburg, Tennessee, April 6 - 11*, 2003.

[124] W. Li, X. Wu, D. Zhang, G. Su, W. Tian, and S. Qiu, "Preliminary study of coupling CFD code FLUENT and system code RELAP5," *Annals of Nuclear Energy*, vol. 73, pp. 96 – 107, 2014.

[125] D. L. Aumiller, E. T. Tomlinson, and R. C. Bauer, "A coupled RELAP5-3D/CFD Methology with a Proof-of-Principle Calculation," *Prodeedings of the 2000 International RELAP5 Users Seminar, Jackson Hole, Wyoming, September 12 - 14*, 2000.

[126] S. B. Rodríguez, "Using the Coupled MELCOR-RELAP5 Codes for Simulation of the Edward's Pipe," Sandia National Laboratories (SNL), SAND2002-2828C, Tech. Rep., 2002.

[127] M. F. Young, K. K. Murata, V. J. Romero, R. O. Gauntt, and G. E. Rochau, "Advanced Nuclear Energy Analysis Technology," Sandia National Laboratories (SNL), SAND 2004-2124, Tech. Rep., 2004.

[128] M. A. Christon, "GILA User's Manual," Sandia National Laboratories (SNL), Albuquerque, NM, SAND2003-1781, Tech. Rep., 2003.

[129] J. R. Travis, P. Royl, J. Xiao, G. A. Necker, and R. Redlinger, "GASFLOW-II: A Computational Fluid Dynamics Code for Gases, Aerosols and Combustion," Karlsruhe Institute for Technology (KIT),, Tech. Rep., 2011.

[130] T. Szabó, F. Kretzschmar, and T. Schulenberg, "Coupling of MELCOR and GASFLOW for more accurate calculation of the hydrogen distribution during severe accidents," *Proceedings of the Annual Meeting on Nuclear Technology 2012, Stuttgart, Germany, May 22 - 24*, 2012.

[131] P. Dietrich, F. Kretzschmar, and A. Miassoedov, "Verbesserungen von Lower Head Modellen in MELCOR - Validierung gekoppelter Modelle durch Simulation der LIVE-Anlage," Karlsruhe Institute of Technology (KIT), Technischer Fachbericht zum Meilenstein M23 des Verbundvorhabens WASA-BOSS, Tech. Rep., 2015.

[132] B. R. Sehgal and Z. L. Yang, "On the experiments with simulant materials at KTH," Royal Institute of Technology, SAM-ECOSTAR-P11/001, Tech. Rep., 2001.

[133] D. J. Rogers and G. J. Janz, "Melting-crystallization and premelting properties of sodium nitrate-potassium nitrate. enthalpies and heat capacities," *Journal of Chemical and Engineering Data*, vol. 27, no. 4, pp. 424 – 428, 1982.

[134] M. M. Kenisarin, "High-temperature phase change materials for thermal energy storage," *Renewable and Sustainable Energy Reviews*, vol. 14, no. 3, pp. 955 – 970, 2010.

[135] T. Jriri, J. Rogez, C. Bergman, and J. Mathieu, "Thermodynamic study of the condensed phases of {NaNO3}, {KNO3} and {CsNO3} and their transitions," *Thermochimica Acta*, vol. 266, pp. 147 – 161, 1995, transition Phenomena in Condensed Matter.

[136] M. Kawakami, K. Suzuki, S. Yokoyama, and T. Takenaka, "Heat capacity measurement of molten NaNO3-NaNO2-KNO3 by drop calorimetry," *Proceedings of the 7th International Conference on Molten Slags Fluxes and Salts, The South African Institute of Mining and Metallury*, 2004.

[137] X. Cheng and U. Müller, "Review on Critical Heat Flux in Water Cooled Reactors," Forschungszentrum Karlsruhe, Institute of Nuclear and Energy Technologies, FZKA 6825, Tech. Rep., 2003.

[138] D. D. Hall and I. Mudawar, "Critical heat flux (CHF) for water flow in tubes: I. Compilation and assessment of world CHF data," *International Journal of Heat and Mass Transfer*, vol. 43, no. 14, pp. 2573 – 2604, 2000.